Glaciers and the Polar Environment

Edited by Masaki Kanao,
Danilo Godone and Niccolò Dematteis

Published in London, United Kingdom

IntechOpen

Supporting open minds since 2005

Glaciers and the Polar Environment
http://dx.doi.org/10.5772/intechopen.87447
Edited by Masaki Kanao, Danilo Godone and Niccolò Dematteis

Contributors
Niccolò Dematteis, Daniele Giordan, Danilo Godone, Valerio Segor, Fabrizio Troilo, Davide Fugazza, Antonella Senese, Roberto Sergio Azzoni, Carlo D'Agata, Daniele Cat Berro, Luca Mercalli, Claudio Smiraglia, Guglielmina Adele Diolaiuti, Fabiano Ventura, Mauro Gobbi, Valeria Lencioni, Muhammad Saifullah, Shiyin Liu, Muhammad Adnan, Muhammad Ashraf, Muhammad Zaman, Sarfraz Hashim, Muhammad Sher, Luis A. Pastene, Yoshihiro Fujise, Jair Putzke, Migdat Hodzic, Ivan Kennedy, Michela Giustiniani, Umberta Tinivella, Rasik Ravindra, Mahesh Badanal, Rahul Mohan, Antonio Batista Pereira

First published in London, United Kingdom, 2021 by IntechOpen
IntechOpen is the global imprint of INTECHOPEN LIMITED, registered in England and Wales, registration number: 11086078, 5 Princes Gate Court, London, SW7 2QJ, United Kingdom
Printed in Croatia

British Library Cataloguing-in-Publication Data
A catalogue record for this book is available from the British Library

Additional hard and PDF copies can be obtained from orders@intechopen.com

Glaciers and the Polar Environment
Edited by Masaki Kanao, Danilo Godone and Niccolò Dematteis
p. cm.
Print ISBN 978-1-83962-592-3
Online ISBN 978-1-83962-593-0
eBook (PDF) ISBN 978-1-83962-594-7

For EU product safety concerns:
IN TECH d.o.o., Prolaz Marije Krucifikse Kozulić 3, 51000 Rijeka, Croatia, info@intechopen.com or visit our website at intechopen.com.

Meet the editors

Dr. Masaki Kanao obtained his Ph.D. from Kyoto University. He is currently working at the National Institute of Polar Research in Tokyo. He is chiefly interested in the Earth's structure and evolution from geoscience studies. Polar regions, both in Arctic and Antarctic, have been investigated by geophysical investigations particularly by passive and active seismic sources. He is also interested in the present Earth's dynamics and tectonics of the continental lithosphere. The Antarctic continent, as a member of the past Gondwana super-continent, has been the main target to reveal lithospheric evolution history. Recently, inter-disciplinary studies in terms of glacial earthquakes, cryoseismic events in Greenland and Antarctica have been focusing on involving environmental changes associated with global warming. These investigations in polar regions have been contributing to the development of all kinds of global Earth sciences.

Dr. Danilo Godone holds a Ph.D. in "Agriculture, Forest and Food Sciences", with his doctorate's topic being the cryosphere's phenomena monitoring by innovative geomatic methodologies. Currently he is a Research Scientist in the Geohazard Monitoring Group (CNR IRPI), studying geomatic contribution in natural hazard monitoring and analysis. His main research interests are landslides, glaciers and, more generally, natural disasters. During his activities, he has developed skills in GIS, R programming, and land surveying with UAVs, GNSS, and LiDAR. He was also involved in several scientific expeditions in the Alps, Nepal, Chilean Andes, and Patagonia. He is a member of NATRISK - Research Centre on Natural Risks in Mountain and Hilly Environments, in Turin University.

Dr. Niccolò Dematteis received his M.S. degree in Physics and he obtained his Ph.D. in Earth Sciences at Pavia University in 2020. He is involved in proximal remote sensing of glaciers and other gravitational processes. His activities concern the measurement of surface deformations using active and passive sensors, like radar interferometers and optical imagery, and data integration techniques. His research interests include glacier monitoring, glacial hazards, and data processing implementation. He is currently affiliated with the Geo-hazards Monitoring Group and works in Turin, Italy.

Contents

Preface

Glaciers and polar regions provide important clues to understanding the past and present status of the Earth system, as well as to predict future forms of our planet.

In particular, Antarctica, composed of an ice-covered continent in its center and surrounded by the Southern Ocean, has been gradually investigated during the last half century by all kinds of scientific branches; bioscience, physical sciences, geosciences, oceanography, environmental studies, together with technological components. Antarctica is now affected by remarkable changes in its temperature and sea-ice extent, mass loss of ice-sheets, variations in marine and terrestrial ecosystems including human activities. Since the most exciting initiative in the polar regions was the International Polar Year (IPY) in 2007-2008, conducted as the 50th anniversary of the International Geophysical Year (IGY 1957-1958). The initiative greatly enhanced the exchange of ideas across nations and scientific disciplines to unveil the status and changes of planet Earth. This kind of inter-disciplinary exchange helps us understand and address grand challenges such as rapid environmental change and its impact on society. In these regards, this book partially aims to compile the achievements of involved projects by the IPY and post era, especially focusing on surface environmental variations associated with climate change.

On the other hand, mountain glaciers are primary sources of freshwater for the population and they sustain the local economies in terms of water supply for agriculture, energy production, and tourism activities. They are suffering due to global warming at a large extent and many low-altitude glaciers are expected to disappear within the next decades. The rapid reduction of the glaciated areas threatens to enhance potential draughts and to modify the distribution, frequency, and magnitude of glacier-related hazards. Moreover, glacial ecosystems and biodiversity might be strongly damaged. For these reasons, the glacier environment has to be carefully studied and monitored, in order to evaluate their possible recent evolution and to implement strategies of resilience, adaptation, and mitigation.

This book covers topics on recent developments of all kinds of scientific research involving glaciers and Antarctica, in the context of currently on-going processes in the extreme environment in polar regions.

Masaki Kanao
Associate Professor,
National Institute of Polar Research,
Tokyo, Japan

Danilo Godone and Niccolò Dematteis
Research Institute for Geo-Hydrogeological Protection,
National Research Council,
Turin, Italy

Chapter 1

Gas Hydrates in Antarctica

Michela Giustiniani and Umberta Tinivella

Abstract

Few potential distributing areas of gas hydrates have been recognized in literature in Antarctica: the South Shetland continental margin, the Weddell Sea, the Ross Sea continental margin and the Wilkes Land continental margin. The most studied part of Antarctica from gas hydrate point of view is the South Shetland margin, where an important gas hydrate reservoir was well studied with the main purpose to determine the relationship between hydrate stability and environment effects, including climate change. In fact, the climate signals are particularly amplified in transition zones such as the peri-Antarctic regions, suggesting that the monitoring of hydrate system is desirable in order to detect potential hydrate dissociation as predicted by recent modeling offshore Antarctic Peninsula. The main seismic indicator of the gas hydrate presence, the bottom simulating reflector, was recorded in few parts of Antarctica, but in some cases it was associated to opal A/CT transition. The other areas need further studies and measurements in order to confirm or refuse the gas hydrate presence.

Keywords: gas hydrate, BSR, Antarctic Peninsula, climate change, opal A/CT

1. Introduction

Gas hydrate is a solid component (clathrates) composed of water and natural gas of low molecular weight (mainly methane), forming under particular condition of low temperature, high pressure, and proper gas concentration [1]. Pressure and temperature define the stability field of gas hydrate, which is affected by gas mixture and pore-fluids composition (salinity). Moreover, the presence of only a small percentage of higher hydrocarbons (such as ethane and propane) shifts the phase boundary to higher temperature (at constant pressure). Generally, hydrates accumulate anywhere in the ocean-bottom sediments where water depth exceeds about 400 m (**Figure 1**). In Polar Regions, in presence of sub-seawater permafrost, the hydrate could be stable at shallower water as demonstrated recently by [2, 3]. Very deep (abyssal) sediments are generally not thought to house hydrates in large quantities due to the lack of high biologic productivity (necessary to produce the organic matter that is converted to methane) and rapid sedimentation rates (necessary to bury the organic matter), both necessary for hydrate formation on the continental shelves. The conditions for gas hydrate stability are verified also in seawater, but gas concentration is always not sufficient for their formations.

Gas hydrates were discovered in 1810 by Humphry Davy [4] and, since then, they became the interest of scientific and engineering research studies. In fact, the stability of methane hydrates on the sea floor has several implications (i.e., "in [5, 6]"). First, they are considered a huge energy resource (i.e., "in [7]"). Second, natural and anthropogenic disturbances may cause their destabilization causing the

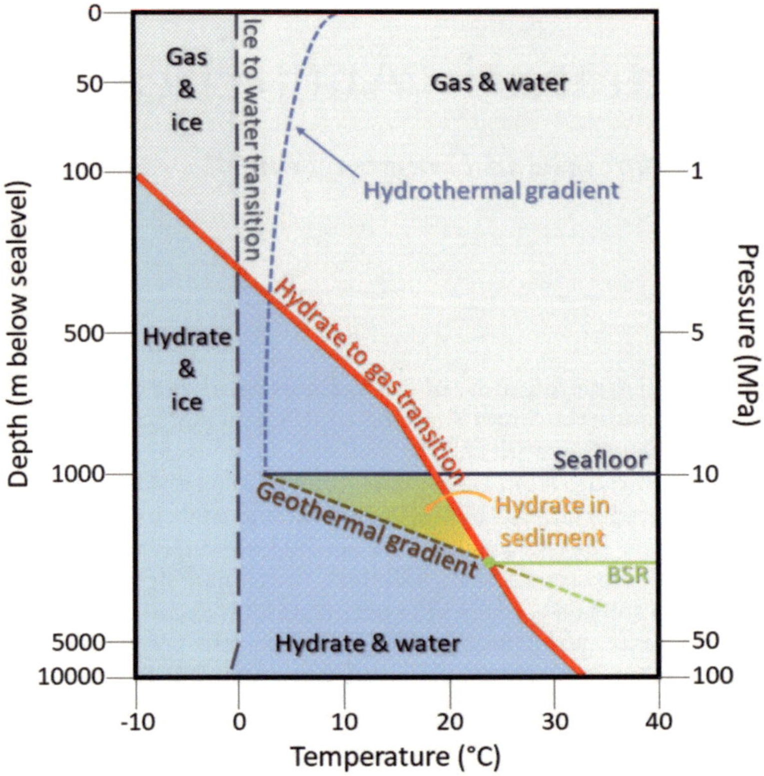

Figure 1.
Schematic diagram of the gas hydrate stability zone in marine environment.

release of huge amounts of fluids (gas and water) and affecting slope stability (i.e., "in [8]"). Finally, methane is an effective greenhouse gas (26 times more powerful than carbon dioxide), and large methane releases may be the cause of sudden episodes of climatic warming in the geologic past (i.e., "in [9]"). Some authors suggested that gas hydrate dissociation influenced significantly climate changes in the late Quaternary period (i.e., "in [10]"). The Clathrate Gun Hypothesis (i.e., "in [11]") suggests that past increases in water temperatures near the seafloor may have induced such a large-scale dissociation, with the methane spike and isotopic anomalies reflected in polar ice cores and in benthic foraminifera [12]. [13] suggested that methane would oxidize fairly quickly in the atmosphere, but could cause enough warming that other mechanisms (for example, release of carbon dioxide from carbonate rocks and decaying biomass) could keep the temperatures elevated.

The relationship between gas hydrate and climate change is of great importance in Polar Regions, where the climate signal is amplified. When pressure and temperature at the sea bottom change (eustatic and climatic changes, respectively), the thickness and the depth of the gas hydrate stability zone change accordingly (i.e., "in [14, 15]"). The study of gas hydrates and the parameters, that control their stability field, allow reconstructing the climatic changes in the past, studying the present processes, and formulating predictions. During the glaciations, the consequent sea level drop produces a rising of the base of the stability field of gas hydrates. This change produces the release of remarkable quantities of methane in the water column, and a sensible continental slope instability, which may cause slides, and, in turn, occasionally tsunami waves. In the other hand, during the interglacial period, the sea level rise and the consequent heating produce an overdeepening of the base of the gas hydrate stability and a progressive accumulation of methane within the gas hydrate zone. Therefore, the climatic changes greatly influence the amounts of methane present in the gas hydrate zone: the release of this gas in the atmosphere during the glaciations influences the interglacial phases, while the decrease of methane content in the atmosphere during the interglacial phases again contributes to the global temperature lowering [16].

Seismic data analysis allow recognizing the presence of gas hydrate in marine environments, because the phase transition (from solid above, to fluid and gasses, below) of interstitial water and gas mixture produces a strong reflection, called Bottom Simulating Reflector (BSR) that simulates the sea bottom and presents a phase reversal with respect to the seafloor reflection. The BSR was firstly discovered and associated to gas hydrate presence in marine sediments in the western Gulf of Mexico, off the northern coasts of Colombia and Panama, and along the Pacific Coast of Central America from Panama to Acapulco by [17]. Successively, in marine environment the BSR was detected along continental margins (both active and passive) and in proximity of mud volcanoes (i.e., "in [18]").

Once thought to be devoid of life, the ice-covered parts of Antarctica are now known to be a reservoir of metabolically active microbial cells and organic carbon [19]. The potential for methanogenic archaea to support the degradation of organic carbon to methane beneath the ice, however, has not yet been evaluated. No data exist forrates of methanogenesis in sub-Antarctic marine sediments. [20] presented experimental data from subglacial environments, similar to Antarctica, that demonstrate the potential for overridden organic matter beneath glacial systems to produce methane. They also numerically simulated the accumulation of methane hydrate in Antarctic sedimentary basins and show that pressure/temperature conditions favor methane hydrate formation down to sediment depths of about 300 meters in West Antarctica and 700 meters in East Antarctica. Moreover, [20] calculated that the sub-Antarctic hydrate inventory could be of the same order of magnitude as that of recent estimates made for Arctic permafrost, suggesting that he Antarctic Ice Sheet may be an important component of the global methane budget, with the potential to act as a positive feedback on climate warming during ice-sheet wastage.

The gas hydrates accumulated in the Antarctic margins could be inferred from geophysical and geochemical evidences, such as BSR on the seismic profile, as already mentioned, high concentrations of methane and organic carbon and abnormal varieties of salinity, chlorinity and sulfate of pore waters in boring sediment samples of Deep Sea Drilling Project (DSDP) and Ocean Drilling Program (ODP) sites. Few potential distributing areas of gas hydrates have been recognized in literature: the South Shetland continental margin, the Weddell Sea, the Ross Sea continental margin and the Wilkes Land continental margin (**Figure 2**).

In Antarctic margins, there are advantageous geological conditions for the formation and accumulation of gas hydrates according to analysis of the reservoir conditions, including gas source, sedimentation, heat flow, temperature, pressure and tectonic conditions, etc. In fact, the modeling of the theoretical base of the gas hydrate stability points out that there is considerable potential resource of gas hydrate in the Antarctic margins. In particular, [21, 22] modeled the gas hydrate distribution in the South Shetland Margin based on geophysical data, while [23] reconstructed the theoretical depth of the BSR in the Ross Sea based on the literature data.

The South Shetland margin (SSM, offshore Antarctic Peninsula) is the most studied part of Antarctica from gas hydrate point of view. In this area, an important gas hydrate reservoir was discovered and was well studied in the recent years with the main purpose to determine the relationship between hydrate stability and environment effects, including climate change.

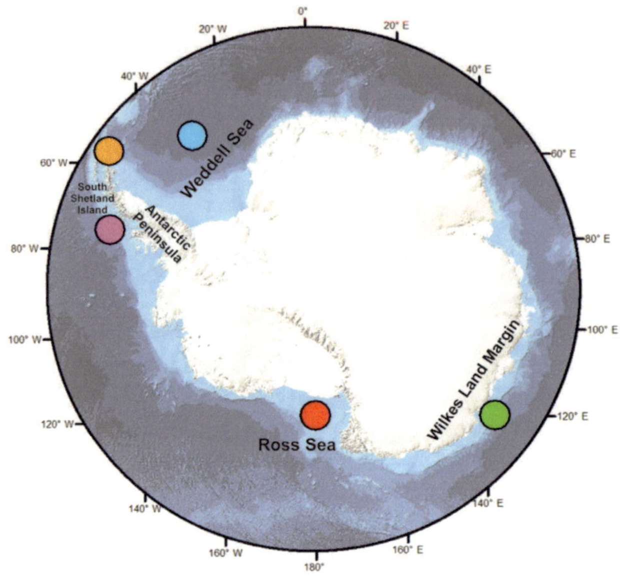

Figure 2.
Map of Antarctica showing the potential areas, indicated by points, where there are indication of gas hydrates presence in literature.

2. The South Shetland Margin (the Antarctic Peninsula)

As mentioned before, the Polar Regions and, in particular, the transition zones, such as the Antarctic Peninsula, are strongly affected by the climate signals. For this reason, several studies are focused on characterizing the South Shetland Margin and the gas hydrate reservoir here present. In the following, we resume the main results related to the geophysical studies.

2.1 Geological setting

The SSM is located in the northeastern tip of the Pacific margin of the Antarctic Peninsula, characterized by the subduction of the Antarctic and the former Phoenix plates beneath the South Shetland micro-continental block. Along the continental margin, a trench-accretionary prism-fore-arc basin sequence can be recognized (i.e., "in [24, 25]"). The Phoenix plate started to subduct beneath the Antarctic plate from late Paleozoic time [26] and progressed from the southwest to the northeast along the margin. Active spreading at the Antarctic Phoenix ridge ceased about 4 Ma ago [27], when the last ridge-crest segment of the Phoenix plate reached the south margin of the Hero Fracture Zone. The subduction process is presently believed to have taken place as a result of sinking and roll-back of the oceanic plate coupled with the extension of the Bransfield Strait marginal basin (i.e., "in [24, 27]"). The Phoenix plate is from about the Shackleton Fracture Zone to the northeastern side, while by the Hero Fracture Zone to the southwestern side, which intersect the continental lithosphere.

2.2 Geophysical data

Different Antarctic expeditions have taken place off the Antarctic Peninsula, in order to verify the existence of a potential gas hydrate reservoir and to reconstruct the tectonic setting of the margin. In this region, the presence of the diffused and discontinuous BRS was discovered during the Italian Antarctic cruses of 1989–1990, 1996/1997, and 2004–2005 onboard the R/V OGS-Explora within the framework of a research program supported by the Italian National Antarctic Program (PNRA; i.e., **Figure 3**; [28–33]. During the first leg, only seismic data were acquired, while during the last two legs Ocean Bottom Seismometer (OBS) and other geophysical data were acquired.

The multidisciplinary dataset, including multibeam bathymetry, seismic profiles (multichannel seismic and OBS data and chirp), and two gravity cores clearly shows active mud volcanism sustained by hydrocarbon venting in the region. The multibeam bathymetric data was indispensable to reconstruct the seabed, covering an area of about 5500 km^2 slides and fluid expulsion related to gas hydrate dissociation or the presence of faults were detected by chirp sub-bottom profiles. The two gravity cores recover 1.07 m and 2.98 m of sediment. Several laboratory measurements were performed on these cores, such as the computer aided tomography and the interstitial fluid analyses to detect gas presence.

2.3 Gas hydrate and related features

The bathymetric map of the SSM provides evidence of four mud volcanoes (**Figure 3**; [18, 31]), which are associated to the presence of gas hydrate. This active mud volcanism might be favored by the reactivation of pre-existing faults and weakness zones because of the regional extensional tectonics of the South Shetland trench and margin [34], the adjacent Bransfield Strait back-arc basin [35], and the complex tectonic interaction at the Elephant Island triple junction [36].

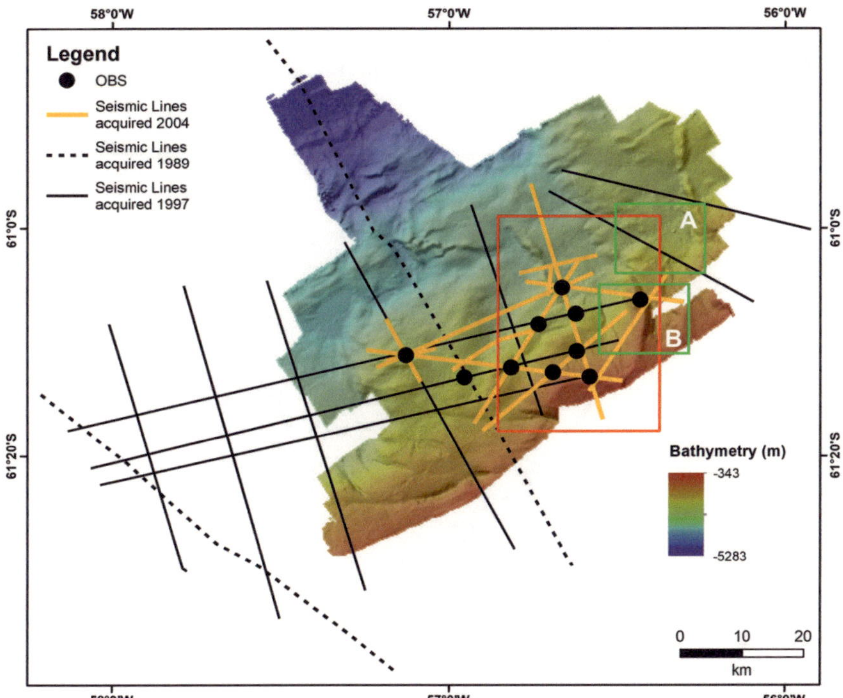

Figure 3.
Map of the investigated area with the location of the acquired data during the surveys. Red rectangle indicates the seafloor reflectivity reported in ***Figure 5****. Green rectangles indicate the position of the zoom reported in* ***Figure 6****.*

The chirp data (location in **Figure 4**) confirm the presence of a few slides, which are probably related to gas hydrate dissociation and several fluid expulsion sites, probably related to active mud volcanism sustained by a hydrocarbon reservoir. Note that, as expected, low values of seafloor reflectivity are identified in correspondence of these features (**Figure 5**). Moreover, the bathymetry shows the presence of ancient slides that could be linked to gas hydrates (**Figure 6**).

Fluid analyses made to the gravity cores revealed the presence of several hydrocarbon gases, such as methane, ethane, propane, butane, pentane and hexane, and traces of aromatic hydrocarbons of >C12 carbon chain length, suggesting a thermogenic origin of the gas.

A strong and continuous BSR was recognized through the analysis of the seismic lines and OBS data, allowing the detection of a large gas hydrate reservoir on the SSM. [33] reported an example of seismic and OBS data. The elastic properties of the different layers across the BSR were modeled by using Tinivella theoretical models in order to quantify the concentrations of gas hydrates and free gas in the pore space [37, 38].

Poisson's ratio near the OBS location was evaluated by the joint inversion of compressional and shear wave arrivals in the vertical and horizontal components of OBS data. Useful information about physical properties of marine sediments in areas where no well data are available were obtained through Amplitude Versus Offset (AVO) analysis and OBS data [29, 38]. In detail, the sediments seems not cemented by the presence of hydrate (due to AVO behavior) and the free gas below the BSR seems uniformly distributed in the pore space (due to the low Poisson ratio) and not in overpressure condition (due to low P-converted wave amplitude; [39]).

Analysis of geophysical data evidences that the accumulation of fluids within sediments is strictly related to tectonics features, such as faults and folds. The concept of hydrate porosity (HP), which is directly related to the fluid content, was introduced in order to better understand the relationship between gas hydrate presence and geological features. HP represents the difference between the

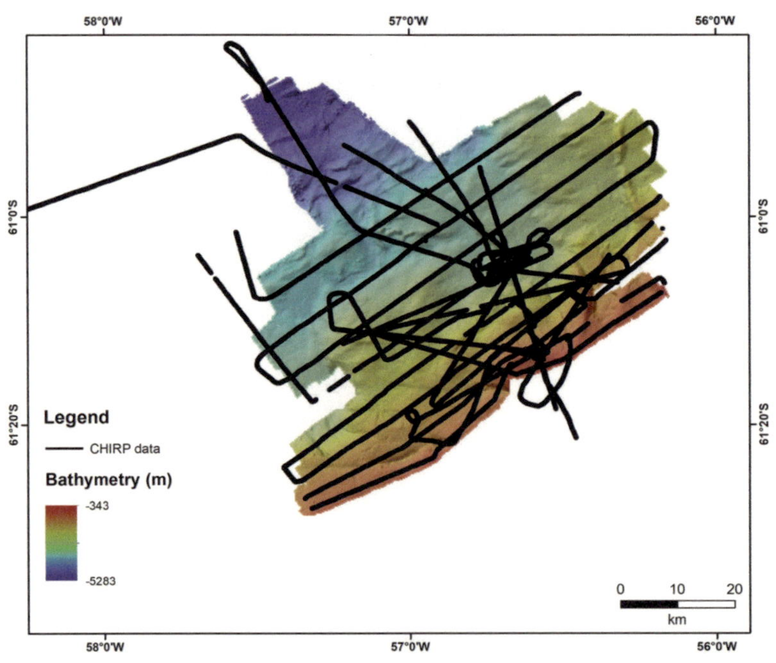

Figure 4.
Map of the investigated area with the location of the CHIRP data.

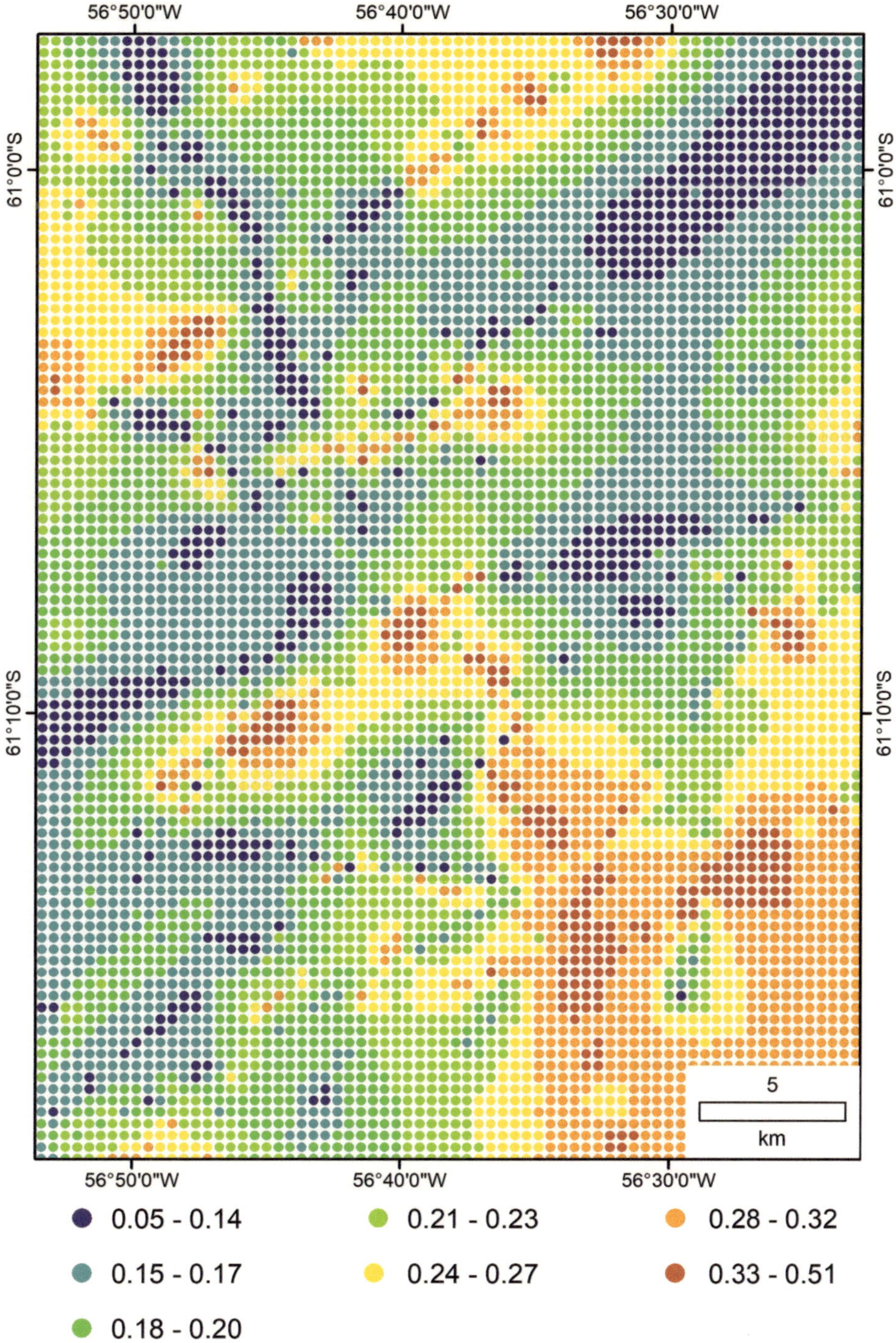

Figure 5.
Map of the seafloor reflectivity extracted from the CHIRP DATA. The location of the area is reports in ***Figure 3****.*

reference porosity (i.e., the porosity without gas hydrate) and the effective porosity (i.e., the porosity reduced by the gas hydrate presence; [40]). The detailed analysis of the reservoir revealed a close relationship between HP, and consequently gas hydrate accumulation, and geological features, such as syncline-anticline structures and fractures distribution within sediments [40]. In particular, a relationship is

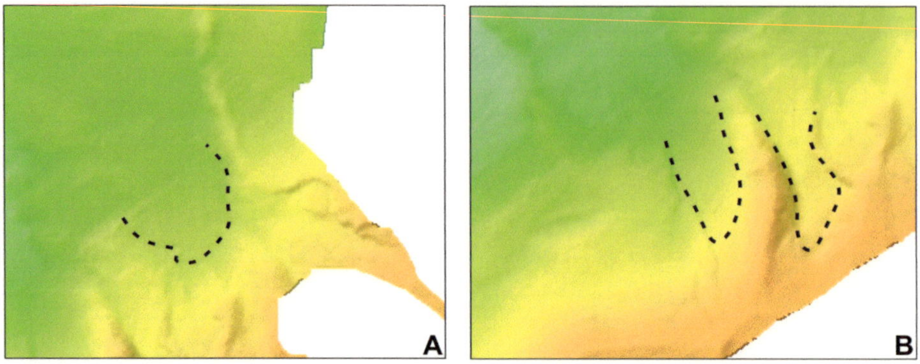

Figure 6.
*Example of ancient slides highlighted with a black dash lines on the bathymetric data. The position of the two zoom is reported in **Figure 3**. The bathymetric data scale is reported in **Figure 3**.*

underlined between the HP values and the distance from the hinge of the anticline: the HP increases toward the limbs of anticline. The microfracturing model supports the idea that the synclines favors the hydrate accumulation above the BSR, while the anticlines favors the free gas accumulation below the BSR, when important faults acting as preferential path-way for fluids escapes [40].

All available seismic profiles and OBS data were analyzed in order to obtain 2D seismic velocity models, then translated in terms of concentrations of gas hydrate and free gas in the pore space by using Tinivella theoretical models [37, 38]. The jointly interpolation of the 2D models allowed obtaining a 3D model of gas hydrate concentration from the seafloor to the BSR, as shown in **Figure 7**. The total volume of hydrate, estimated in the area (600 km^2) where the interpolation is reliable, is 16×10^9 m^3. The gas hydrate concentration is affected by error estimated equal to about ±25%, as deduced from sensitivity tests and from error analysis related to the interpolation procedure. The estimated amount of gas hydrate can vary in a range of 12×10^9–20×10^9 m^3. Moreover, considering that 1 m^3 of gas hydrate corresponds to about 140 m^3 of free gas in standard conditions, the total free gas trapped in this reservoir ranges between 1.68×10^{12} and 2.8×10^{12} m^3. This estimation does not take into account the free gas contained within pore space below the hydrate layer, so this values could be underestimated [41].

2.4 Modeling of gas hydrate versus climate change

In the past years, scientists are taking an interest in modeling warming-induced hydrate dissociation in the Antarctic region. Over the period 1958 to 2008, the Antarctic Peninsula shows an unusually high rate of warming [42], the strongest of the Southern Hemisphere and one of three strongest on Earth [43]. Predicting future warming in this area is challenging because of the lack of a sound physical mechanism that explains the present regional warming [43], but some models predict that in the 21st century the Antarctic Peninsula may not experience the strongest warming of Antarctica [44]. Ocean warming in West Antarctica is predicted to be of 0.5 ± 0.4° C by 2100, about half of global mean warming, considering the A1B scenario [45], which assumes modest reductions in greenhouse-gas emissions after mid-21st century. A long-term ocean warming similar to that predicted in West Antarctica may be sufficient to trigger dissociation of a shallow hydrate reservoir in the SSM. This hypothesis has been preliminary tested by [22] based on steady-state modeling of the evolution of the base of the hydrate stability zone assuming a 1.4° C increase by the end of the 21st century.

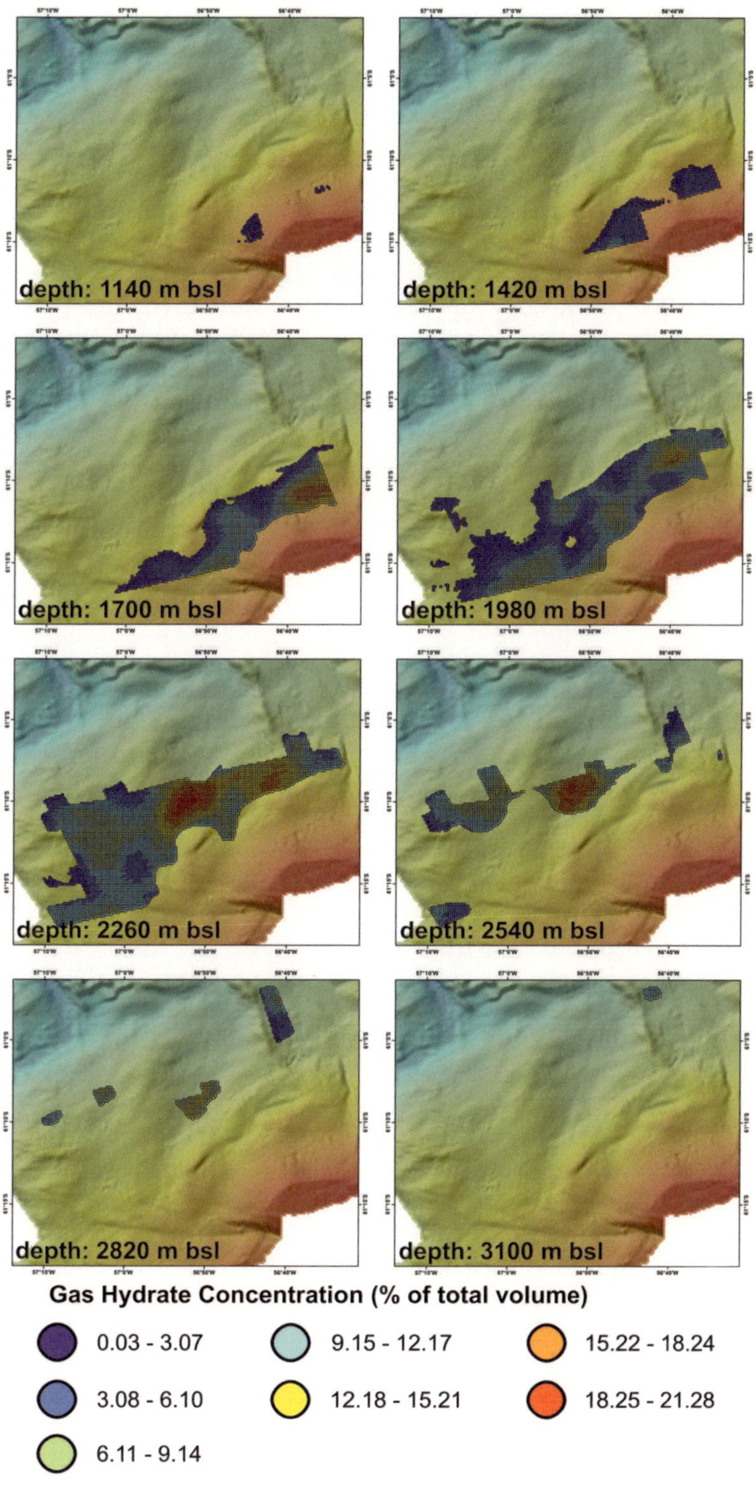

Figure 7.
Map of the gas hydrate concentrations at different depth from seafloor (in meters).

Successively, it was modeled the transient response to ocean warming of the hydrate system in the SSM between 375 and 450 mwd for the period 1958–2100 CE, using constraints in input parameters from seismic observations [22, 46].

TOUGH-HYDRATE (T-H) code [47] was employ for the modeling, with past temperatures given by the US National Oceanographic Data Center and two future temperature scenarios given by extrapolation of the temperature trends over the periods 1960–2010 and 1980–2010. The result of the transient modeling shows that methane emissions may occur at water depths between 375 m and 425 m if the future seabed temperatures follow a similar trend to that over the period 1980 to 2010 of 0.0238° C y-1), while emissions would not occur with a seabed warming rate an order of magnitude smaller [46]. Hydrate dissociation would initiate at the top of the hydrate layer, and the overpressure generated would not be sufficient to cause, by itself, shallow slope failures or shallow vertical fractures over the 21st century. Hydrate-sourced methane emissions at 375 mwd would start at ca. 2028 and may extend to deeper waters at an average rate of 0.91 mwd y^{-1}. Over the 21st century, the potential amount of dissociated methane liberated to the ocean may be between 1.06 and 1.21–10^3 mol y^{-1} per meter along the margin [46]. This modeling underlines that the SSM is one of the key areas to observe and understand the effects of warming-induced hydrate dissociation in the Southern Hemisphere during the coming decades [46].

3. Weddell Sea

The Weddell Sea is considered a potential area for gas hydrate accumulation (i.e., "in [48]"), even if a clear indication of hydrate presence is missed. It is important to underline that, in this part of Antarctica, acquisition of data was very difficult in the past due to presence of ice shelves. Only in the last years, the extraordinary rapid climate warming, which is occurring in the northern tip of the Antarctic Peninsula [49, 50], caused the reduction of land ice along West Antarctica and the ice shelves destruction in the surrounding seas (i.e., "in [51, 52]"). In north-western of Weddell Sea, [53] detected the presence of gaseous hydrocarbons (from methane to n-pentane) in the seabed sediments and the bubbling of methane suggesting the presence of gas accumulations in the substrate of the NW Weddell Sea. They observed a release of methane from the frozen ocean substrate adjacent to Seymour Island, linked to climate instability during Late Cenozoic, when vast areas of the Antarctic continental shelf were flooded during the marine transgression that occurred *c*. 18,000 years ago, after the Last Glacial Maximum. The heat flow from the sea to the marine substrate, now flooded, would have destabilized frozen gas accumulations, which were originally formed into terrestrial permafrost during the Last Glacial Maximum, similarly to what would have happened in the Arctic [53].

Seismic data acquired in 1985 over the south-eastern continental shelf and the margin of the South Orkney microcontinent as a site survey for ODP Leg 113 [54], show a BSR lying at 500–800 ms. The widespread cause of the reflection was interpreted as a break-up unconformity associated with the 25–30 Ma opening of the Jane Basin to the east [55]. In places, the detected BSR cuts across beddings, and in this case this physical boundary may be either depositional or of secondary origin related to the diagenesis of biogenic silica, possibly combined with a major variation of the detrital input [56]. So, also in this case, the BSR is not produced by gas hydrate and free gas presence, suggesting that a careful analysis of seismic data is necessary before to interpret a BSR as the base of gas hydrate stability zone.

So, potentially, in this area, all conditions to have gas hydrate are verified, even if the small amount of data acquired cannot confirm or reject this hypothesis.

4. Wilkes Land margin

[57] inferred gas hydrates to be present in sediments offshore Wilkes Land. A multichannel seismic-reflection survey revealed a reflector showing the characteristics of a BSR [58]: (1) the reflector is at a depth consistent with the pressure/temperature stability field of gas hydrate, and (2) the reflector shows a reversal of polarity. Unfortunately, a third criterion that the subbottom depth of the reflector increases with increasing depth of water is not met, possibly because of oceanward increasing geothermal gradients as in the case of the inferred gas hydrate off Japan [59]. In addition, other seismic data acquired in 1993 by Japan National Oil Corporation in collaboration with Geological Survey of Japan [60] revealed possible BSRs that could be associated to zones with gas hydrate.

Clearly, additional data are required to confirm and eventually characterize the presence and the distribution of gas hydrate in the Wilkes Land Margin.

5. The Ross Sea

In the last two decades, the Ross Sea Embayment area has been considered a laboratory of growing interest for the reconstruction of the past Antarctic environment, the onset of Antarctic Eocene-Paleocene glaciation, climates studies, the understanding of the tectonic deformation and global sea level changes that all have driven glacial history [23].

The main Ross Sea elongated N-S sedimentary troughs, such as the Victoria Land Basin, the Drigalsky Basin, the northern continuation of the Northern Basin, the Central Basin, the Joides Basin and the Eastern Basin are bounded by basement highs and morphological banks. They were formed during the late Cretaceous major rifting phase and later during the Cenozoic, while a widespread igneous activity affected the West Antarctic Rift System ([23] and references therein).

The potential presence of gas hydrate is supported by the identification of hydrocarbons in the Ross Sea area. In fact, in the central and eastern Ross Sea, the cored Miocene muddy sediments at the DSDP sites showed high contents of total hydrocarbon gas (mainly methane; [61]). In the western Ross Sea, analysis of sediments from gravity cores showed the presence of hydrocarbon gases with low concentrations of methane [62] and in the McMurdo Sound, both CIROS-1 and MSSTS-1 wells, detected small amount of organic carbon [63, 64]. Moreover, [65] supposed the presence of a BSR, an indirect indication of the presence of gas hydrate, on a seismic line, located in the Victoria Land Basin. Moreover, in the same part of the Ross Sea, an extensive field of pockmarks at 450–500 m depth and unusual flat-topped seafloor mounds was identified on a detailed multibeam dataset [66, 67]. One hypothesis discussed by the authors is that these features may be carbonate banks because of their proximity to the inferred subsurface gas hydrate, although their preferred interpretation is that the features are of volcanic origin.

At this stage, to confirm or refuse the presence of gas hydrate in the Ross Sea more measurements are necessary. Recently, [23] performed a modeling of the base of gas hydrate stability based on the steady-state approach by using literature data, such as bathymetric and well data, sea bottom temperature, a variable geothermal gradient and assuming that the natural gas is methane, in order to identify the areas where the gas hydrate stability is verified The modeling was performed in the whole Ross Sea (**Figure 8**).

The results from the modeling suggested that depth and distribution of the base of the gas hydrate stability zone are correlated with the bathymetry. In fact,

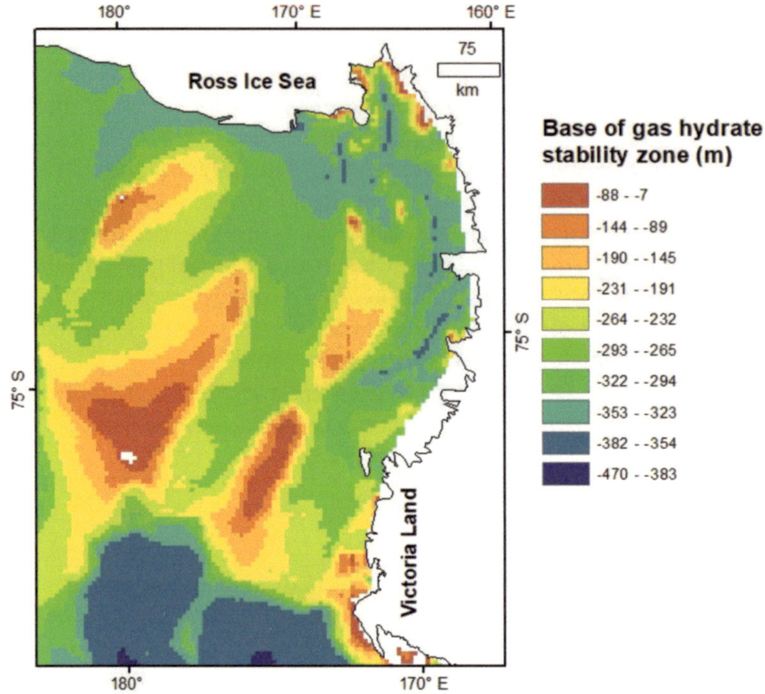

Figure 8.
Distribution of the base of the gas hydrate stability zone from the seafloor (in meters). The geothermal gradient is supposed to be equal to 49°C/km (modified after [23]).

in proximity of the banks, the gas hydrate stability zone results display thickness less than 100 m. On the other hand, the thickness of the gas hydrate stability zone increases in proximity of the basins to values exceeding 400 m related to bathymetry increase and seafloor temperature decrease. Moreover, the existence and dynamics of the gas hydrate distribution is strictly related to the existence and evolution of the shallow geological and geomorphological features below the sea floor, as suggested in the past by several authors. So, the presence of some geological and geomorphological features are in agreement with the gas hydrate presence in this part of Antarctica [23].

6. Opal-A/opal-CT phase boundary

Silica diagenesis consists of precipitation from an initial amorphous phase (opal A) to an intermediate phase (opal CT) and finally to the final form with quartz crystallization. The presence of large amounts of biogenic silica in marine sediments can affect their physical properties [68]. In fact, diagenetic alteration of biogenic opal-A to opal-CT causes a drastic reduction of porosity (about 20 vol% according to [69]), which contributes to sediment consolidation at depth.

It is possible to recognize the passage from one phase to another on seismic profiles because of the presence of a high amplitude reflector, produced by a positive impedance contrast between the overlying silica-rich sediments and the underlying sediments in which biogenic silica is dissolved. This reflector simulates the seafloor morphology, so it is still called BSR. This BSR is different from the hydrate-related BSR, as well documented in literature (i.e., "in [70]"). The positive polarity, the depth, no noticeable drop of frequency, and compressional velocity below the

BSR suggest the absence of the gas hydrate and the diagenetic-related origin of the observed BSR. Moreover, diagenesis-related BSR have a constant depth below seafloor or even decreasing sub-bottom depth with increasing water depth due to earlier opal transition at greater pressures.

6.1 Pacific margin

In the Pacific margin of the Antarctic Peninsula, several seismic lines where acquired with the main purpose to study sediment drift presented in the northwest part. There lines were analyzed in detail in order to extract information about the petrophysical properties relevant to seismic stratigraphy studies in the continental shelf and rise [71, 72]. In particular, a seismic line showed the presence of an anomalous reflector, interpreted as a BSR [73]. Borehole data are also available, thanks to the ODP Leg 178 [74].

In order to understand the nature of the observed BSR, [73] performed a detailed study concluding that the BSR observed in the seismic line is due to opal-A/opal-CT phase boundary and not to the gas hydrate presence. Moreover, they attempted a quantitative estimation of biogenic silica content within marine sediments using seismic reflection and physical properties data across the silica diagenesis-induced BSR. The estimated biogenic silica content increases with depth and reaches a maximum of 23.3 wt % above the BSR. Such quantifications are of prime importance for submarine slope stability assessment as the deep seated transformation of biogenic silica from opal-A to opal-CT is able to trigger slope instability not only at local scale but also at regional scale, as previously shown by [69, 75].

7. Conclusions

The most important area from gas hydrate point of view in Antarctica is the South Shetland Margin where a huge hydrate reservoir is present; it is very well documented in literature thanks to several geophysical acquisition legs performed. The analyses of geophysical data allows concluding that the accumulation of fluids within sediments is strictly related with tectonics features, such as faults and folds, revealing a close relationship between gas hydrate accumulation and geological features. Moreover, the hydrocarbons trapped and detected in the sediment cores may indicate the existence of deeper reserves, confirming that the BSR should be considered as an indicator of conventional deep reservoir. Finally, due to the warming measured in this part of Antarctica, a monitoring of the evolution of the gas hydrate reservoirs offshore Antarctic Peninsula is required for an environmental in-depth analysis.

The main seismic indicator of the gas hydrate presence, the BSR, was recorded in few parts of Antarctica (Pacific and Atlantic margin of Antarctic Peninsula, Wilkes Land margin), but it was associated to opal A/CT transition. The other potential areas for gas hydrate presence (Ross Sea, Weddell Sea and Wilkes Land Margin) needs further measurements in order to confirm or refuse the hypothesis of their presence.

Acknowledgements

We wish to thank all colleagues that contributed to improve the knowledge of the gas hydrate in Antarctica. In particular, in alphabetic order: Flavio Accaino (OGS, Italy), Daniela Accettella (OGS, Italy), Angelo Camerlenghi (OGS, Italy),

Emanuele Lodolo (OGS, Italy), Maria Filomena Loreto (former OGS, Italy, now Istituto di Scienze Marine, CNR, Italy), Hector Marin-Moreno (former OGS, Italy, now Norwegian Geotechnical Institute, Norway), Jong Kuk Hong (Korea Polar Research Institute, Korea), Xuewei Liu (China University of Geosciences, China), Cristina Neagu (former OGS, Italy), Sha Song (former OGS, Italy, now Chang'an University, China). The Authors have been partially supported by the Programma Nazionale di Ricerche in Antartide (PNRA), the Ministry of Foreign Affairs, the TALENTS FVG Programme - European Social Fund, and the Ministry of Education, Universities and Research under the grant for Italian participation in the activities related to the international infrastructure Partnership for Advanced Computing in Europe (PRACE).

Conflict of interest

The authors declare no conflict of interest.

Author details

Michela Giustiniani[1*] and Umberta Tinivella[2]

1 National Institute of Oceanography and Applied Geophysics - OGS, Sgonico, Italy

2 National Institute of Oceanography and Applied Geophysics - OGS, Udine, Italy

*Address all correspondence to: mgiustiniani@inogs.it

References

[1] Sloan E.D. Jr. 1998. Clathrate Hydrates of Natural Gases. 2nd ed. Marcel Dekker; 1998. 705 p.

[2] Tinivella U., Giustiniani M. Gas hydrate stability zone in shallow Arctic Ocean in presence of sub-sea permafrost. Rendiconti Lincei. 2016; 27: 163-171. DOI: 10.1007/s12210-016-0520-z

[3] Tinivella U., Giustiniani M., Marín-moreno, H. A quick-look method for initial evaluation of gas hydrate stability below subaqueous permafrost. Geosciences. 2019; 9: Article number 329. DOI: 10.3390/geosciences9080329

[4] Davy H. On a combination of oxymuriatic gas and oxygen gas: Philosophical Transactions of the Royal Society; 1811.155 pp.

[5] Henriet J. P., Mienert J. Gas hydrates: relevance to world margin stability and climate change. Geol Soc Spec Publ. 1998; 137.

[6] Liu X., Flemings P. B. Dynamic multiphase flow model of hydrate formation in marine sediments. Journal of Geophysical Researcher. 2007; 112 (B03101). DOI: 10.1029/2005JB004227.

[7] Collett T. Energy resource potential of natural gas hydrates: AAPG Bulletin. 2002; 86: 1971-1992.

[8] Mienert J., Bünz S., Guidard S., Vanneste M., Berndt C. Ocean bottom seismometer investigations in the Ormen Lange area offshore mid-Norway provide evidence for shallow gas layers in subsurface sediments. Marine and Petroleum Geology. 2005; 22: 287-297. DOI: 10.1016/j.marpetgeo.2004.10.020

[9] Waite W. F., Santamarina J. C., Cortes D. D., Dugan B., Espinoza D. N., Germaine J. , Jang J., Jung J. W., Kneafsey T. J., Shin H., Soga K., Winters W. J., Yun T.-S. Physical properties of hydrate-bearing sediments, Rev. Geophys. 2009; 47: RG4003. doi:10.1029/2008RG000279.

[10] Kennett J. P., Cannariato K. G., Hendy I. L., Behl R. J. Carbon isotopic evidence for methane hydrate stability during Quaternary Interstadials. Science. 2007; 288: 128-133. DOI: 10.1126/science.288.5463.128.

[11] Kennett J. P., Cannariato K. G., Hendy L. L., Behl R. J. Methane Hydrates in Quaternary Climate Change: The Clathrate Gun Hypothesis. Spec. Publ. 2002, 54, AGU, Washington, D. C.

[12] Reagan M.T., Moridis G.J. Oceanic gas hydrate instability and dissociation under climate change scenarios. Geophysical Research Letters. 2007; 34 (22). DOI: 10.1029/2007GL031671

[13] Haq B.U.. Natural gas hydrates: searching for the long-term climatic and slope-stability records. In: Henriet J.P., Mienert J., editors. Gas Hydrates: Relevance to World Margin Stability and Climate Change. Special Publications: London, Geological Society; 1998. 137, 303-318.

[14] Fyke J.G., Weaver A.J. The Effect of Potential Future Climate Change on the Marine Methane Hydrate Stability Zone. Journal of Climate. 2006; 19: 5903-5917. DOI: 10.2307/26259334

[15] Alessandrini G., Tinivella U., Giustiniani M., Vargas-Cordero I., Castellaro S. Potential instability of gas hydrates along the chilean margin due to ocean warming. Geosciences. 2019; 9: Article number 234. DOI: 10.3390/geosciences9050234

[16] MacDonald G.T. Role of methane clathrates in past and future climates. Climatic Change. 1990; 16: 247-281.

[17] Shipley T. H., Houston M. H., Buffler R. T., et al. Seismic reflection evidence for widespread occurrence of possible gas-hydrate horizons on continental slopes and rises. AAPG Bull. 1979; 63: 2204-2213.

[18] Tinivella U., Giustiniani M. An Overview of Mud Volcanoes Associated to Gas Hydrate System. Updates in volcanology-new advances in understanding volcanic systems. 2012. DOI: 10.5772/51270

[19] Priscu J.C., S. T., Tulaczyk S., Studinger M., Kennicutt II M.C., Christner B. C., Foreman C. M. Antarctic subglacial water: origin, evolution and ecology. In: Vincent W. F., Laybourn-Parry J., editors. Polar Lakes and Rivers. Oxford University Press; 2009.

[20] Wadham J. L., Arndt S., Tulaczyk S., Stibal M., Tranter M., Telling J., Lis G. P., Lawson E., Ridgwell A., Dubnick A., Sharp M. J., Anesio A. M., Butler C. E. H. Potential methane reservoirs beneath Antarctica. Nature. 2012; 488: 633-637. DOI:10.1038/nature11374

[21] Loreto, M.F., Tinivella U., Accaino F., Giustiniani M. Correlation between geological structures and gas hydrate amount offshore the South Shetland Island — preliminary results. Advances in Geosciences. 2010; 18: 223-231. DOI: 10.1142/9789812838148_0014

[22] Tinivella U., Giustiniani M., Accetella D. BSR versus Climate Change and Slides. Journal of Geological Researc. 2011; Article ID 390547. doi:10.1155/2011/390547

[23] Giustiniani M., Tinivella U., Sauli C., Della Vedova B. Distribution of the gas hydrate stability zone in the Ross Sea, Antarctica. Andean Geology. 2018; 45(1): 78-86. DOI: 10.5027/andgeov45n1-2989

[24] Kim H., Cho M., Lee, J. Low-pressure thermal metamorphism of volcanic rocks in the Barton Peninsula. King George Island. Antarctica. In: Kim Y. et al., editors. The fourth Seoul International Symposium on Antarctica Science - Geology of the South Shetland Islands. Seoul: Korea Ocean Research and Development Institute. 1995. 25-27.

[25] Maldonado. A., Larter R., Aldaya F. Forearc tectonic evolution of the South Shetland margin, Antarctic Peninsula. Tectonics. 1994; 3:1345-1370.

[26] Pankhurst R.J. The Paleozoic and Andean magmatic arcs of West Antarctica and southern South America. In: Kay S.M., Rapela C.W., editors. Plutonism from Antarctica to Alaska. Geological Society of American; 1990. Boulder, CO, USA, 1990; pp. 1-7.

[27] Larter R.D., Barker P.F. Effects of ridge crest-trench interaction on Antarctic-Phoenix spreading: Forces on a young subducting plate. J. Geophys. Res. 1991; 96: 19583-19607.

[28] Lodolo E., Camerlenghi A., Brancolini G. A Bottom Simulating Reflector on The South Shetland Margin, Antarctic peninsula. Antarctic Science. 1993; 5(2): 207-210. DOI: 10.1017/S0954102093000264

[29] Tinivella U., Accaino F. Compressional velocity structure and Poisson's ratio in marine sediments with gas hydrate and free gas by inversion of reflected and refracted seismic data (South Shetland Islands, Antarctica). Mar. Geol. 2000; 164, 13-27.

[30] Tinivella U., Carcione M. Estimation of gas-hydrate concentration and free-gas saturation from log and seismic data. The Leading Edge. 2001. 20.

[31] Tinivella U., Accaino F., Della Vedova, B. Gas hydrates and active mud volcanism on the South Shetland

continental margin, Antarctic Peninsula. Geo-Mar. Lett. 2008; 28: 97-106.

[32] Tinivella U., Loreto M.F., Accaino F. Regional versus detailed velocity analysis to quantify hydrate and free gas in marine sediments: the south shetland margin case study. Soc. Geol. Lond. Special Publ. 2009; 319: 103-119.

[33] Song, S., Tinivella U., Giustiniani M., Singhroha S., Bunz S., Cassiani G. OBS data analysis to quantify gas hydrate and free gas in the South Shetland margin (Antarctica). Energies. 2018; 12(1): Article number 3290.

[34] Loreto M.F., Tinivella U., Ranero C.R. Evidence for fluid circulation, overpressure and tectonic style along the Southern Chilean margin. Tectonophysics. 2006; 429 (3-4): 183-200

[35] Barker D. H., et al. Backarc basin evolution and cordilleran orogenesis: Insights from new ocean-bottom seismograph refraction profiling in Bransfield Strait. Antarctica. Geology. 2003; 31: 107-110. doi:10.1130/0091-7613(2003)031<0107:BBEACO>2.0.CO;2.

[36] Klepeis K.A., Lawver L.A. Tectonics of the Antarctic–Scotia plate boundary near Elephant and Clarence Islands, West Antarctica. J Geophys Res. 1996; 101:20211-20231

[37] Tinivella U. A method to estimate gas hydrate and free gas concentrations in marine sediments. Bollettino di Geofisica Teorica ed Applicata. 1999; 40: 19-30.

[38] Tinivella U. The seismic response to overpressure versus gas hydrate and free gas concentration. Journal Seism. Explor. 2002; 11: 283-305.

[39] Tinivella U., Giustiniani M. Variations in BSR depth due to gas hydrate stability versus pore pressure. Global and Planetary Change. 2013; 100: 119-128

[40] Loreto M.F., Tinivella U. Gas hydrate versus geological features: The South Shetland case study. Marine and Petroleum Geology. 2012; 36(1): 164-171

[41] Loreto, M.F., Tinivella U., Accaino F., Giustiniani M. Offshore Antarctic Peninsula gas hydrate reservoir characterization by geophysical data analysis. Energies. 2011; 4(1): 39-56

[42] Mulvaney R., Abram N.J., Hindmarsh R.C., Arrowsmith C., Fleet L., Triest J., Sime L.C., Alemany O., Foord S. Recent Antarctic Peninsula warming relative to Holocene climate and ice-shelf history. Nature. 2012; 489: 141-144.

[43] Vaughan D.G., Marshall G.J., Connolley W.M., Parkinson C., Mulvaney R., Hodgson D.A., King J.C., Pudsey C.J., Turner J. Recent rapid regional climate warming on the Antarctic Peninsula. Climatic Change. 2003; 60: 243-274.

[44] Chapman W.L., Walsh J.E. 2007. A synthesis of Antarctic temperatures. Journal of Climate. 2007; 20: 4096-4117

[45] Yin J., Overpeck J.T., Griffies S.M., Hu A., Russell J.L., Stouffer R.J. Different magnitudes of projected subsurface ocean warming around Greenland and Antarctica. Nature Geosciences. 2011; 4: 524-528.

[46] Marín-Moreno H., Giustiniani M., Tinivella U. The potential response of the hydrate reservoir in the South Shetland Margin, Antarctic Peninsula, to ocean warming over the 21st century. Polar Research. 2015; 34:1, 27443. DOI: 10.3402/polar.v34.27443

[47] Moridis G.J., Kowalsky M.B., Pruess K. TOUGH-HYDRATE v1.2 user's manual: a code for the simulation of system behavior in hydrate-bearing geological media. Berkeley, CA: Lawrence Berkeley National Laboratory, University of California. 2012

[48] Terzariol M., Park J., Castro G. M., Santamarina J. C. Methane hydrate-bearing sediments: Pore habit and implications. Marine and Petroleum Geology. 2020; 116: 104302. Doi:10.1016/j.marpetgeo.2020.104302

[49] Davies B.J., Carrivick J.L., Glasser N.F., Hambrey M.J., Smellie J.L. Variable glacier response to atmospheric warming, northern Antarctic Peninsula, 1988-2009. Cryosphere. 2012; 6(5): 1031-1048.

[50] Glasser N. F., Scambos T. A., Bohlander J. A., Truffer M., Pettit E. C., Davies B. J. From ice-shelf tributary to tidewater glacier: continued rapid glacier recession, acceleration and thinning of R¨ohss Glacier following the 1995 collapse of the Prince Gustav Ice Shelf on the Antarctic Peninsula. Journal of Glaciology. 2011; 57(203): 397-406.

[51] Rignot E. Changes in West Antarctic ice stream dynamics observed with ALOS PALSAR data," Geophysical Research Letters. 2008; 35(12).

[52] Rignot E., Bamber J. L., Van Den Broeke M. R., Davis C., Li Y., Jan van de Berg W., van Meijgaard E. Recent Antarctic ice mass loss from radar interferometry and regional climate modelling. Nature Geoscience. 2008; 1(2): 106-110.

[53] del Valle R. A., Yermolin E., Chiarandini J., Sanchez Granel A., Lusky J.C. Methane at the NW of Weddell Sea, Antarctica. Journal of Geological Research 2007; vol. 2017: Article ID 5952916. Doi:10.1155/2017/5952916.

[54] Barker P.F., Kennett J.P., et al., 1990. Proc. ODP, Sci. Results, 113: College Station, TX (Ocean Drilling Program). Doi:10.2973/odp.proc.sr.113.1990.

[55] Barker P. R, Barber P. L., King E. C. An early Miocene ridge crest-trench collision on the South Scotia Ridge near 36°W. Tectonophysics. 1984; 102:315-332.

[56] Lonsdale M. J. The relationship between silica diagenesis, methane, and seismic reflections on the South Orkney Microcontinent. In: Barker P. F., Kennett J. P., editors. Proceedings of the Ocean Drilling Program, Scientific Results. Ocean Drilling Program. Texas A&M University, College Station, TX. 1990; Vol. 113, pp. 27-37.

[57] Kvenvolden K.A., Golan-Bac M., Rapp J.B. Hydrocarbon geochemistry of sediments offshore from Antarctica: wilkes Land continental margin. CPCEMR Earth Sci. 197; Series 5A, 205-213.

[58] Eittreim S.L., Uchupi E. Marine geological and geophysical investigations of Antarctic continental margin. U.S. Geological survey, Circ., 935, 1-12.

[59] Yamano M., Uyeda S., Aoki Y., Shipley T.H. Estimates of heat flow derived from gas hydrates: Geology. 1982; 10: 339-343.

[60] Tsumuraya Y., Tanahashi M., Saki T., Machihara T., Asakura N. Preliminary report of the marine geophysical and geological surveys off Wilkes Land, Antarctica, in 1983-1984. Mem. Natl. Inst. Polar Res. Spec. Issue. 1985; 37, 48-62.

[61] McIver R.D. Hydrocarbon gases in canned core samples from leg 28 sites 271, 272 and 273, Ross Sea. In: Initial Reports of the Deep Sea Drilling Project. 1975; 28: 815-817. Washington. Doi: 10.2973/dsdp.proc.28.133.

[62] Rapp J.B., Kvenvolden K.A., Golan-BacM. Hydrocarbongeochemistry of sediments offshore from Antarctica. In: Cooper A.K., Davey F.J.; editors. The Antarctic Continental Margin, Geology and Geophysics of the Western Ross Sea. Circum-Pacific Council for Energy and Mineral Resources, Earth Science Series 5B; 1987: 217-224. Houston.

[63] Collen J.D., Xinghua Y., Collier R.J., Johnston J.H. 1989. Hydrocarbon source rock potential and organic maturation. In: Barrett P.J., editor. Antarctic Cenozoic history from the CIROS-1 Drillhole, McMurdo Sound. D.S.I.R. Bull. 1989; 245: 223-230. New Zealand.

[64] White P. Downhole logging. In: Barrett P.J., editor. Antarctic Cenozoic history from the CIROS-1 Drillhole, McMurdo Sound. Department of Scientific and Industrial Research, Bulletin. 1989; 245: 7-14. Wellington.

[65] Geletti R., Busetti M. A double bottom simulating reflector in the western Ross Sea, Antarctica. Journal of Geophysical Research: Solid Earth. 2011; 116 (B4): B04101. Doi: 10.1029/2010JB007864.

[66] Lawver L.A., Davis M.B., Wilson T.J., Shipboard Scientific Party. Neotectonic and other features of the Victoria Land Basin, Antarctica, interpreted from multibeam bathymetry data. In: International Symposium on Antarctic Earth Sciences, No. 10, Extended Abstract 017: 1-4. Santa Bárbara. 2007

[67] Lawver L., Lee J., Kim Y., Davey F. Flat-topped mounds in western Ross Sea: Carbonate mounds or subglacial volcanic features? Geosphere. 2012; 8 (3): 645-653. Doi: 10.1130/GES00766.1.

[68] Bryant W.R., Rack F.R. Consolidation characteristics of Weddell Sea sediments: results of ODP Leg 113. In: Proc. scientific results, ODP, Leg 113, Weddell Sea, Antarctica; 1990. pp. 211-223. Doi: 10.2973/odp.proc.sr.113.173.1990.

[69] Volpi V., Camerlenghi A., Hillenbrand, C.-D., Rebesco, M., Ivaldi, R. Deep sea sediment consolidation through the glacial history of the Pacific margin of the Antarctica Peninsula. Terra Antartica Reports. 2003; 9: 51-56.

[70] Berndt C., Bunz S., Clayton T., Mienert J., Saunders M. Seismic character of bottom simulating reflectors: examples from the mid-Norwegian margin. Marine and Petroleum Geology. 2004; 21(6):723-733

[71] Tinivella U., Camerlenghi A., Rebesco M. Data report: Seismic velocity analysis on the continental shelf transect, ODP Leg 178, Antarctic Peninsula. In: Proceedings of the Ocean Drilling Program: Scientific Results; 2002: 178:1-25

[72] Volpi V., Camerlenghi A., Moerz T., Corubolo P., Rebesco M., Tinivella U. Data report: Physical properties relevant to seismic stratigraphic studies, continental rise sites 1095, 1096, and 1101, ODP leg 178, Antarctic Peninsula. In: Proceedings of the Ocean Drilling Program: Scientific Results. 2002; 178:1-36

[73] Neagu R.C., Tinivella U., Volpi V., Rebesco M., Camerlenghi A. Estimation of biogenic silica contents in marine sediments using seismic and well log data: Sediment Drift 7, Antarctica. International Journal of Earth Sciences. 2009; 98(4), 839-848

[74] Barker P. F., Camerlenghi A., Acton G. D. Leg 178 Summary. In: Proc ODP Init. Rep. 1999; 178:60.

[75] Davies R.J., Clark I.R. Submarine slope failure primed and triggered by silica and its diagenesis. Basin Res. 2006; 18:339-350

Chapter 2

Geomorphological Insight of Some Ice Free Areas of Eastern Antarctica

Rasik Ravindra, Badanal Siddaiah Mahesh and Rahul Mohan

Abstract

The Schirmacher Oasis and Larsemann Hills are among the few significant ice free areas of East Antarctica that are conspicuous due to presence of more than a hundred melt water lakes each, preserving the signatures of climatic variation and deglaciation history since Last Glacial Maximum (19 to 24 ky BP) and beyond. There are evidences, recorded in the lake sediments of low lying Larsemann Hills, of marine transgression due to variation in sea level, isostatic upliftment and close vicinity of the Hills to the marine environment. The Schirmacher Oasis, on the other hand has preserved various landforms-both erosional and depositional-typical of a periglacial environment along with proglacial lakes (incorporating signals of ice-sheet dynamics) and epishelf lakes (signatures of marine influence) .

Keywords: geomorphological evolution, Schirmacher and Larsemann Hills, Paleoclimate

1. Introduction

The Antarctic continent, comprised of two distinct physiographic and tectonic domains i.e. the West Antarctica and the East Antarctica is divided into two unequal parts by a 3500 km long Transantarctic Mountain chain extending across the continent between the Ross and the Weddell Sea (**Figure 1**). The inhospitable climate, inaccessible terrain conditions with 98% of the terrain being covered by a thick apron of ice, the scanty outcrops are the best alternative (if not the only) to peep into the Continent's geological history. The rock outcrops are exposed in discontinuous mountain chains, along the coastal fringes of Antarctic Peninsula, West Antarctica, the Dronning Maud land, Enderby Land, Princess Elizabeth Land, Wilkes Land and Victoria Land in the east Antarctic Sector, apart from the Trans-Antarctic Mountains. (**Figure 1**). The interior of the Antarctic mainland is entirely ice covered and rise as ice plateau, attaining maximum height of around 3233 m above mean sea level (m.s.l.).

2. Ice free areas of East Antarctica

The East Antarctic coast is marked by a discontinuous mountain chain that can be traced intermittently all along the coast from 75^0 45′ E longitudes to 15^0

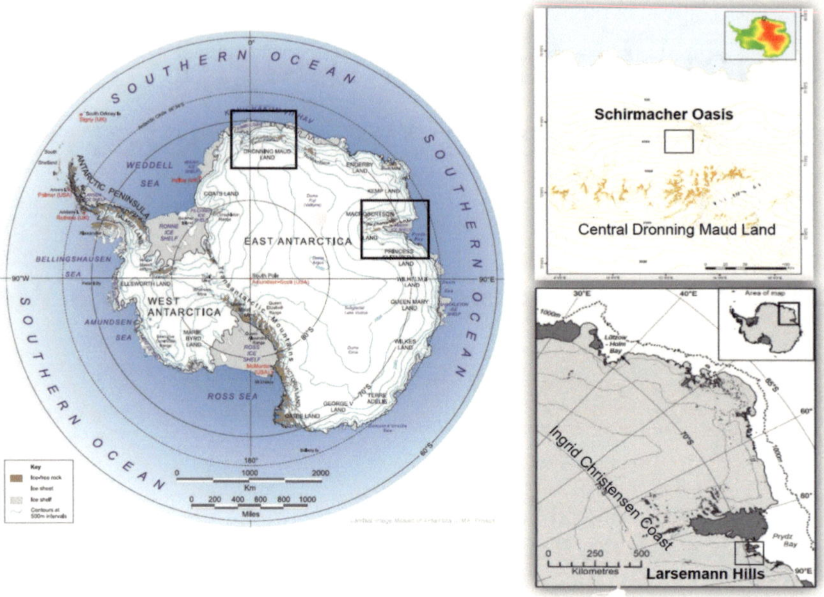

Figure 1.
Map of Antarctica the Southern Ocean (Landsat image mosaic of Antarctica (http://lima.nasa.gov/pdf/A3_overview.pdf), with inset showing locations of Schirmacher and Larsemann Hills.

West longitudes, running nearly parallel to the coast line. There are some low lying, ice free areas in the coastal Antarctica such as the Schirmacher Oasis, Larsemann Hills, Vestfold Hills, Bunger Hills etc., that have been studied in detail for Late Pleistocene (~ 0.12 My) and Holocene glacial History [1–12]. These oases are distinguished from nunataks by the process of ablation. While most nunataks are located in the accumulation zone of glaciers and are kept free of ice by the strong winds, the oases are separated from the ice sheet by a distinct ablation zone. Schirmacher Oasis and Larsemann Hills, the two areas being discussed here, are such ice free regions that were covered by the ice sheet during the Last Glacial Maxima (LGM) or earlier, but are now exposed due to retreat of ice sheet.

Schirmacher Oasis (approximately 35 km^2 in area) is situated between longitudes 11^0 25′ E and 11^0 55′E and latitudes 70^0 44'S and 70^0 46'S, about 85 km inland of the Princess Astrid coast at the northern fringe of central Dronning Maud land. Larsemann Hills on the other hand, is spread over ~50 Km ^{2} [12] and comprise a group of ice free peninsulas (Broknes, Stornes and Brattnavet) Grovnes promontory and islands (McLeod, Fischer, Sandercock Island etc.) - located south of Prydz Bay at 69"24'S, 76′20'E on the Ingrid Christensen Coast of Princess Elizabeth Land that lie in between the Vestfold Hills and Amery Ice shelf. The two areas (Schirmacher Oasis and Larsemann Hills) are nearly 3000 km apart and experience different scale of environment, severity of climatic conditions and paleoclimate history.

The retreating ice sheet left bare rocks to be exposed to the strong Antarctic winds and other erosional processes. The areas demonstrate subdued topography with strong control of lithology and structure over the landscape. The low lying hills are devoid of horns, arêtes or conical peaks. The flat hills, dominate the landscape. The softer rock material has weathered out giving way to the glacial melt to form scores of lakes in the depressions thus created. The freeze–thaw cycles, frost action and the salt weathering are conspicuous and have resulted in formation of conspicuous landscape typical of periglacial and glacial environment [13].

3. Schirmacher oasis

3.1 Physiography

Schirmacher Oasis and adjoining areas depict contrasting morphological units, viz. (a) the ice shelf to the north, (b) the structural hills, and (c) the continental or polar ice sheet to the south (**Figures 2**, **3** and **5**). The morphological features as seen in the ice and ice dominated regions around Schirmacher constitute an integral part of geomorphology of the region.

The ice shelf extends for about 80 km north of Schirmacher Oasis towards the Southern Ocean and displays a highly rugged and broken undulating upper surface dissected by a number of pressure ridges, crevasses and pods of melt water concentration in the western parts, as compared to a low gradient surface in the eastern sector. The Pressure ridges, formed due to the tidal activity in the sea below, and the obstacle offered by the landmass, are often seen at the contact of ice shelf and continent (**Figures 3** and **5**). A number of melt water channels concentrated close to the hills, in this part, can be noted. The continental ice sheet that encircles the Schirmacher Oasis overrides the bare rocks on its southern side. It has a regional

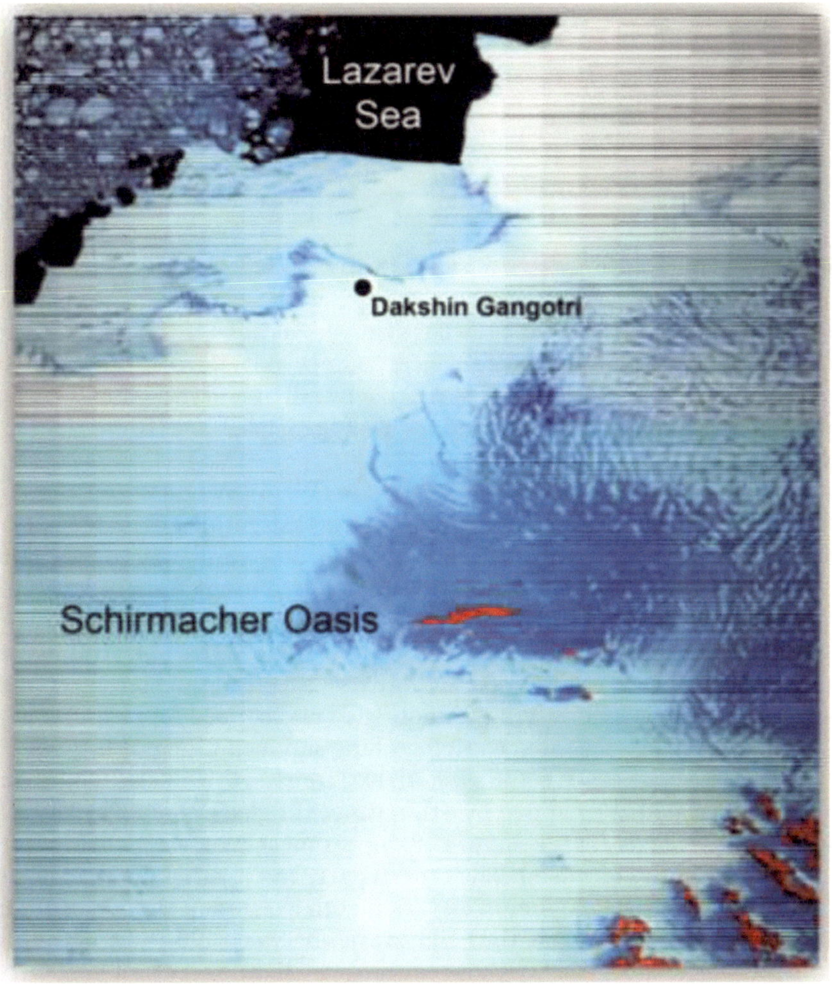

Figure 2.
Map of Schirmacher oasis showing location of Maitri (Indian Research Station) with ice shelf and continental ice.

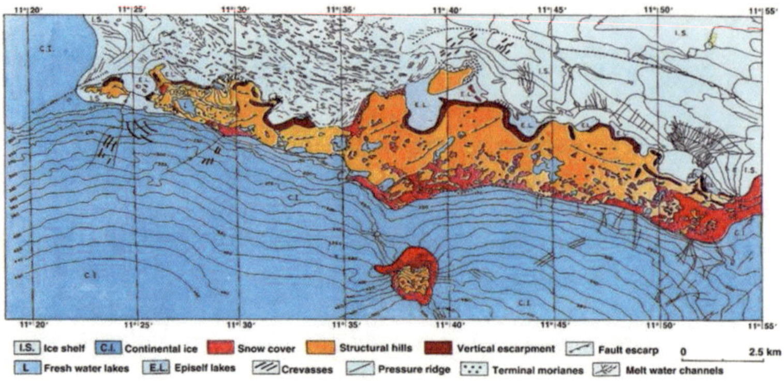

Figure 3.
Geomorphological map of Schirmacher oasis.

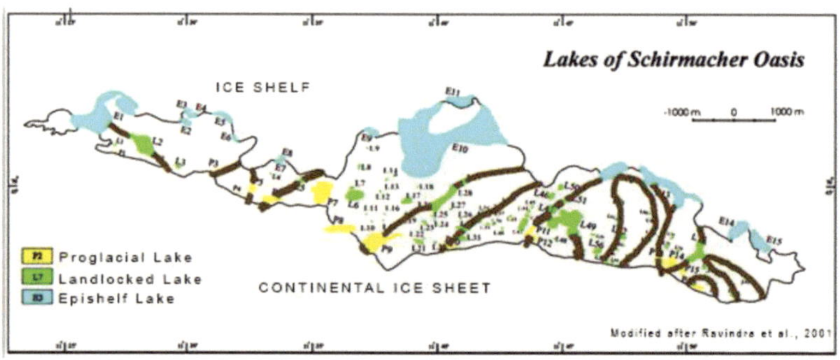

Figure 4.
Distribution of fresh water lakes in Schirmacher oasis and paths of glacier movement.

northerly gradient but locally the flow has swerved because of the nunataks and the land mass of Schirmacher which offered a resistance to its normal flow. The western component of gradient is conspicuous. A broken trail of end moraines is seen at the contact of continental ice-sheet and rocks of Schirmacher at some places in eastern margin.

The rocks of Schirmacher Oasis are aligned in ENE-WSW direction and represent middle Proterozoic sequence of quartzo-feldspathic gneiss, augen gneiss, quartzite and sillimanite garnet gneiss etc. bearing close similarity to the khondalite rocks of eastern and southern India. The rocks exposed in the form of low lying hills extend roughly for about 17 km in length and 3 km in width (at its widest in the central parts), covering an area of ~35 sq. km. The ice-free area exhibits rolling subdued topography with concordant hill tops. At both its eastern and western extremities, the hills are comparatively low lying, smoothened and capped by a thin veneer of glacier boulders. The ice-free area exhibits rolling subdued topography with concordant hill top. At both its eastern and western extremities, the hills are comparatively low lying, smoothened and capped by a thin veneer of glacier boulders The elevation varies from sea level (at the margin of westernmost lake, which has broken the apparent continuity of rocks). The central part, on the other hand, exposes hills that are comparatively higher in elevation with isolated peaks of the order of 212 m and 228 m. On an average, the height varies between 120 and 130 m above m.s.l. The southern end of the Schirmacher range is slightly elevated in comparison to the steep northerly margin. The northern margin is conspicuous by its vertical escarpment almost all along its length (**Figure 3**). The escarpment at places is more than 140 m as in the central part (70° 43′ 30″S,11°42′E). The jagged hills south of Schirmacher Oasis forms a part of the Wohlthat Mountains which rise to elevation 3500 m above m.s.l. leading to Polar Plateau, further south wards (**Figure 2**).

The landscape of Schirmacher Oasis is dotted by more than a hundred lakes of varying dimensions (**Figure 4**). The retreating ice cap vacated the land mass and exposed the rocks of Schirmacher Oasis, scooping out the material from weaker zones in the terrain. There were also blockages created at the mouth of glaciers due to dumping of debris carried by some glaciers. Due to these phenomenon, a number of inland lakes came into existence. Aerial survey over Schirmacher Oasis reveals a definite pattern of concentration of these lakes. Considering the morphological disposition of such lakes and their genesis, these lakes have been grouped in three different classes viz., Proglacial Lakes, Periglacial Lakes (also land locked/inland lakes) and Epishelf Lakes [14]. Proglacial and periglacial lakes dominate in number over the Epishelf lakes and together account for about 87% of the total lakes. The Proglacial Lakes, formed as a result of scouring of the rocks lying at the foot of snout tongues of glaciers, are located at the margin of continental ice sheet and run all along the southern end of Schirmacher Oasis. A NNE–SSW trending lineament cutting across all the three physiographic units viz. continental ice sheet, shelf ice and hard rock, located in the eastern part, defines a prominent fault running for nearly 8.5 km. It is seen as a well-defined crevasse zone in the former two units while in the later unit, it manifests itself in the form of shearing, tight folding, and escarpment demonstrating the structural control over geomorphology.

In the central region, the orientation and location of some inland lakes define a palaeo-channel. The path of the extinct glacier is evident from the U-shaped valley containing sporadic shallow lakes. These lakes have been carved out from structural and lithological weak zones like shears, lineaments, faults etc. (**Figure 3**). The palaeo-paths of the glaciers reconstructed using the evidences of glacial striations, moraine deposits etc., indicate a bimodal direction of the glaciers i.e., NE to ENE in the western and central parts while NNW in the eastern parts (**Figure 4**). This observation is supported by the results obtained by GPS campaign [15] that shows varying magnitude of the horizontal velocities in the range of 1.89–10.88 ma^{-1}. There are a number of epishelf lakes that are located at the northern margin of Schirmacher Oasis which have been described as 'sea bays' (**Figure 6**) as these are connected to sea from beneath and thus respond to tidal waves as is evidenced by pressure ridges. The loci of these epishelf lakes also coincide with the sites where glacial flows must be debauching the ablation material including moraines and melt water as is evidenced by the concavity of the hills and vertical escarpment at these locations.

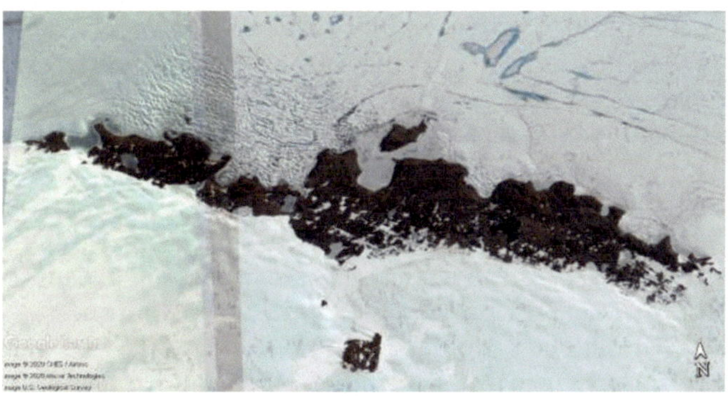

Figure 5.
Aerial photographic mosaic of Schirmacher displaying three physiographic units (A, B and C), Pressure Ridges and disposition of lakes defining lineaments.

Figure 6.
An Epishelf lake (E13) at northern margin of Schirmacher oasis. Note the vertical escarpment at the margin of lake. Continental ice sheet is located overriding part of hill (photo courtesy: Prof. Yusuke Suganuma, NIPR, Japan, SONIC: India-Japan coring expedition).

3.2 Deglaciation history

The Antarctic ice sheet extended across the continental shelf edge, before and during the LGM. The interior surface-elevation of ice sheet did not change significantly, but there was thickening of the ice around the edge of Antarctica. The LGM ice volume accounted for ~120 m of sea level lowering. Abrupt sea level rise occurred at 19,000 calendar years ago [16] following the beginning of Termination/deglaciation and at 14,200 (Melt water Pulse 1A, MWP1A). Detailed records of $\delta^{18}O$ reveal that the last isotopic maximum (LIM, near 18,000 cal years BP) is younger than the LGM as defined by sea level low stand (~ 21 cal ka BP). This suggests early warming of the deep sea, and implies that the deep ocean circulation must have played a key role in the termination of the LGM. During the last glacial cycle (between 19 and 71 cal ka BP), a sizeable portion of high latitude continental shelf was occupied by ice sheets. Ice sheets on the shelf were inherently unstable, being controlled by sea level. Therefore shelf glaciations played a critical role in the dynamics of deglaciation.

The isostatic rebound, a consequential to retreat of ice sheet, has resulted in uplifting of the landmass. Though the exact component of uplift cannot be quantified due to lack of beach features, the morphological evidences such as: a) existence of comparatively higher relief of the structural hills on the northern periphery of the landmass than the central corridor, b) the steep escarpment at the northern margin, and c) the indication of a fault running all along the northern margin give credence to the statement.

The different processes of deposition and erosion under the prevailing periglacial environment have left their imprints on the morphology of Schirmacher Oasis. There was an extensive phase of erosion in operation, during and after the retreat of glacier as evidenced from the erosional features such as a) rolling topography, b) absence of sharp peaks, c) glacial striations and polishing of the rock surfaces and d) the existence of en-echelon pattern of the Roche Moutonees over a large area in the oasis. Features such as block fields, cavernous pits, etc. were formed due to extreme variation in the diurnal temperatures and strong wind erosion. The superimposition of the wind features on the glacial imprints, as seen under electron microscope imply the long period of exposure of the terrain to the weathering processes after the retreat of the ice mass. The depositional features are marked by extensive moraines, terraces, erratic boulders on hill tops and lacustrine deposits. The detailed description of landforms is given by [13].

The existing planar surfaces of the Schirmacher Oasis offer a unique landscape that indicates a pre-Holocene weathering profile. In the absence of the reliable dating, it is not possible to comment with firm conviction if the surface has been a result of Mesozoic weathering profile subsequent to fragmentation and breaking of Gondwanaland supercontinent or a Pleistocene event. However, reported ages of 53.7 ± 8.2 and 51.2 ± 9.4 ka from two sediment cores [17] obtained from the lakes of Schirmacher Oasis could be correlated with the beginning of the oxygen isotopic stage 3 (MIS 3). The surficial glacial till deposit have been dated by him at 30 to 40 cal ka BP [17]. Such an old age does give credence to the hypothesis that Schirmacher was ice free during LGM and before. This is also supported by the studies [18] from other Eastern Antarctic Oases, such as Bunger and Larsemann Hills, believed to be vacated by ice much before LGM.

Two long sediment cores collected from the L-49 have been dated at different depths. The oldest dates obtained from the basal and near basal sections at 168 to 174 cm from the top have been dated at 30,640 years and 32,655 years BP. Cold conditions prevailed in the Schirmacher Oasis from 30,640–21,685 years B.P. having a low sedimentation rate of 0.005 mm/year. Warmer conditions existed between 32,655–30,640 years B.P. with a higher sedimentation rate of 0.015 mm/year. The ^{14}C dates of another core suggested a wet climate between 29,920–28,890 years B.P. with a sedimentation rate of 0.09 mm/year [19]. Study of clay minerals from core samples has led [20] propose that there was a gradual shift in the weathering regime and climate from strongly glacial to fluvio- glacial specially around 42 ka..

Reconstruction of the paleoclimate history from the pollen spores present in the sediment samples of Lake (L-49) by [21] shows that the region witnessed cold and dry climate during 10–9 ka B.P. followed by a long phase of warm and moist climate from 9 to 2.4 ka B.P. Subsequently from 2.4–1 ka B.P. onwards, dry and cold conditions set in the Schirmacher Oasis. However, the climate ultimately turned warm and moist beginning at 1 ka B.P. The sedimentation rate of the fluvio-glacial deposits, especially in the lakes give an indication of the varying paleoclimate. It is evident from the studies that between 8000 and 3500 years BP the climate was warm as compared to the period before and after it so as to yield fast inflow of the sediments in the lakes during this period. The interpretation is in conformity with the palyonological data [22] that infers a warm, humid and warm & humid climate between this time span, on the basis of pollen studies. These alternating phases of climate were made on the basis of dominance of grasses, cosmarium (fresh water algae) etc.

The lake history from 13 ka B.P. to the present has also been attempted by using the magnetic and geochemical properties of seven vertical sediment profiles along an east–west transect in Schirmacher Oasis [23]. Further, based on the results of AMS ^{14}C dates [24], reports that greater parts of Schirmacher was dominated by glaciers from 13 to 12.5 ka B.P and colder conditions prevailed in the Schirmacher Oasis between 13 and 12.5 ka B.P.; ~12–11.5 ka B.P. and 9.5–5 ka B.P. However, due to the onset of warming conditions (~11.5 ka B.P.), the glaciers retreated leading to the formation of five large pro-glacial lakes which are now located on the low lying valleys of the Schirmacher Oasis. Based on the environmental magnetic properties of sediments deposited in Sandy Lake, glacial–interglacial climatic variation was reconstructed for the past 42.5 cal. ka B.P. [23]. Extremely cold periods in the Schirmacher Oasis were recorded during 40.8, 36, 34.51, 29 and, 28.02–21.45 cal. Ka B.P. Relatively warm periods were documented during 38.4–39.2 cal. ka B.P., 33.7–29.8 cal. ka B.P. and 28.5 cal.ka B.P. The Holocene period was characterized by alternating phases of relatively warm (12.55–9.9 cal. ka B.P.and 4.21–~2 cal. ka B.P) and cold (9.21–4.21 cal. Ka B.P. and from ~2 cal. ka B.P. onwards) events. These results are in conformity with results of other studies, as documented above. Further, the geochemical proxies (TC%, TN%, C/N ratios, $\delta^{13}C$ and $\delta\ ^{15}N$) along with the physical proxies (grain size: sand-silt-clay) for three different periglacial lakes viz., Long Lake, Zub Lake and Sandy Lake [1–3] spanning the glacial–interglacial variations (spanning up to 43 cal ka BP). These studies presents the evolution of lake through reconstruction of productivity patterns, source of organic matter and the hydrological processes through grain size variation complimenting the environmental magnetism records from the same lakes [5, 6]. The deglaciation history from the above observation suggest most likely that parts of Schirmacher Oasis were ice-free even during the LGM. This can be supported records of consistency in the continuity of the sedimentary sequences. However, parts of Schirmacher Oasis became ice free during the last deglaciation i.e., Termination 1. Hence, to better understand the deglaciation history of Schirmacher Oasis, the sedimentary records needs to be supplemented by further studies using novel techniques such as cosmogenic dating of rock outcrops and erratic all across Schirmacher Oasis.

4. Larsemann Hills/Prydz Bay Area

The Larsemann Hills represents one of the largest coastal ice free area of Antarctica, located in the Prydz Bay Region on the Ingrid Christensen Coast. The area is comparatively free of ice shelf with hills protruding in the sea as two prominent peninsulas- Broknes and Stornes Peninsula. In between these two, smaller landmasses namely Grovnes and Brattnevet promontory and several smaller islands (McLeod, Fisher, and Bens etc.) dot this region (**Figure 7**). The area exposes Proterozoic felsic/gneissic basement, overlain by a pelitic and psammitic paragniesses rocks, dominated by medium to coarse grained garnet bearing gneisses as compared to Archean gneisses with s crossing cutting mafic dykes found in Vestfold Hills.

4.1 Physiography

The Ingrid Christensen Coast is marked by a zig-zag coast line with sporadic occurrence of low lying hills. Theses hills are over ridden by continental ice sheet

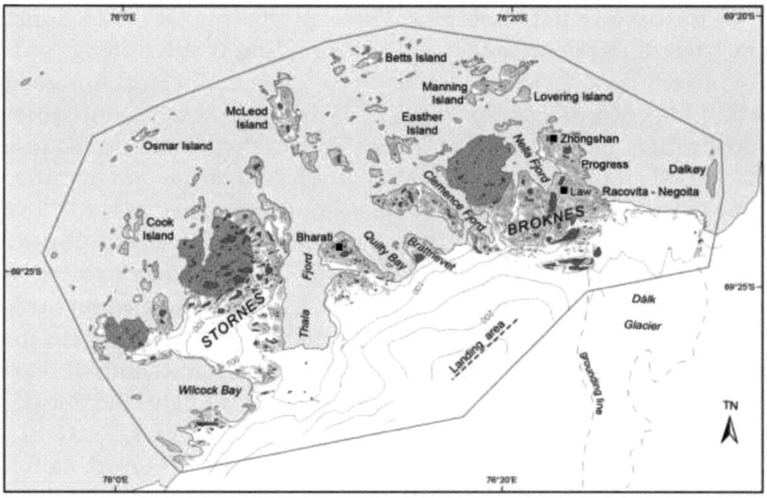

Figure 7.
Map of Larsemann Hills showing major ice free hills. (https://www.antarctica.gov.au/).

towards south exhibiting steep gradient. The Polar Record Glacier, the Polar Times Glacier and the Polarforschung Glacier flow west of Larsemann Hills in the Publication Ice shelf as a part of the Lambert Glacier system. The Publication Ice shelf itself marks a very steep margin east of Polar Record Glacier till it ends abruptly south east of Bolingen Islands. Fjords, such Thala, Quilty and Clemence have cut deep into the ice reaching the ice sheet while Nella fjord has made its way through a valley in the Broknes Peninsula.

The Larsemann Hills exhibit horst and graben structures, signifying a tectonic control in their disposition. The Stornes Peninsula, Grovnes Promontory, Fischer Island and Broknes Peninsula from west to east, are conspicuous as horsts with their straight N-S to NNE–SSW trending margins with Thala fjord, Quilty Bay, Clemence Fjord and Nella Fjord respectively flowing in the complimentary grabens (**Figure 7**). The hills are totally devoid of peaks and show low hummocky and rounded tops with isolated erratic boulders on top (**Figure 8**). The general elevation varies between 50 and 70 m above m.s.l. with some parts approaching ~100 m above m s l. The hills are dissected by vertical valleys demonstrating past fluvio-glacial action. The hill tops show weathered surfaces with pits while the windward sides show cavernous pits at places. Moraines are rare. The landscape is dotted with more than 150 water bodies which have been mapped and numbered systematically by [25]. While most of the lakes are shallow up to 3 m depth, and may be classified as ephemeral ponds, there are some deep lakes such as Lake Progress (34 m) in Broknes Peninsula, Lake Oskar in Stornes Peninsula (18 m) and lake LH 7 (14 m) in Grovnes (Bharati) Promontory. Some lakes are saline in nature due to close vicinity of the ocean and the wind born salt spray and/or excessive evaporation and support a thick biomass mat in upper levels. .

Earlier records [26, 27] have given an account of the post glacial regional climate variability along the East Antarctic coastal margin. A detailed description of the region (ANARE reports) and of the microbial communities inhabiting the lakes of Larsemann Hills are provided elsewhere [28–30]. The minimum age of deglaciation of the islands has been inferred to be late Pleistocene/early Holocene. However, reconstruction of relative seal level (RSL) records [31] have stressed that the presence of marine sediments with radiocarbon ages ranging from 40 to 30 ka BP to the east of Filla Island suggest that deglaciation of some areas could have commenced much earlier. Diatom abundance and fossil pigment records [32, 33] also opine that Larsemann Hill were not fully covered by ice during the LGM and gradually deglaciated between c. 13.5 and 4 ka BP. Relatively wet conditions prevailed between c. 11.5 and 9.5 ka BP in a lake on one of the northern islands in the Larsemann Hills [26]. Basal samples from a sediment core at ~158 cms in Larsemann Hills revealed presence of a marine sediments [34] while two more records based on diatom

Figure 8.
Larsemann Hills. Low hills with hummocky tops near Grovnes promontory. Note the open sea in austral summer and the continental ice sheet marking coast line.

records and geochemical proxies suggest presence of marine sediments during mid-Holocene [3, 35] and while diatoms endemic to sub-Antarctic island corresponding to MIS 5e stage were recorded from Broknes Peninsula (Last Interglacial) [33, 36]. The reduced elevation and planation surfaces must have been carved before the LGM facilitating the marine transgression. Based on the variation of diatom population present in sediments suggest that the influence of seawater got weakened after ~5000 yrs. BP [37] and relative warmer climatic condition was prevalent [37, 38]. Core studies from a lake in Grovnes Promontory by [39] have shown high productivity between ~8.3 to ~6 cal ka BP and that the ice free conditions prevailed around 4 cal ka BP.

5. Conclusions

The geomorphology of Schirmacher Oasis and Larsemann Hills, the two prominent ice free areas present distinct erosional and depositional landforms characteristic of a polar periglacial environment. Both the areas house hundreds of melt water lakes that have preserved the signatures of glacial–interglacial climate variations in the sediments deposited. Paleaoclimate history for the past ~42 cal. ka B.P [5] has revealed wide spread glaciation in LGM with Holocene being characterized by alternating warm and cold phases. Presence of marine sediments were documented from two coastal lakes viz., Pup Lagoon [35] and Heart Lake [3, 35] suggesting isostatic upliftment in the region. The sedimentological data viz.: sediment sorting, composition and the SEM studies on quartz grains from the two areas have shown similar results except the extent and strength of glacial processes. While there was a strong effect of glacial processes on quartz grains from Larsemann Hills, the imprints of glaciofluvial activity were more prominent on quartz grains from Schirmacher Oasis [40].

Acknowledgements

The authors thank the Ministry of Earth Sciences, Government of India and expeditioners of Indian Antarctic Programme. MB and RM would also like to place on record their sincere thanks to Director, NCPOR for the encouragement. This is NCPOR Contribution No **B-5/2020-2021.**

Author details

Rasik Ravindra[1*], Badanal Siddaiah Mahesh[2] and Rahul Mohan[2]

1 National Centre for Polar and Ocean Research, Goa, India

2 National Centre for Polar and Ocean Research (NCPOR), Ministry of Earth Sciences, Headland Sada, Goa, India

*Address all correspondence to: rasikravindra@gmail.com

References

[1] Mahesh B S, Warrier A K, Mohan R, Tiwari M, Babu A, Chandran A, Asthana R and Ravindra R. Response of Long Lake sediments to Antarctic climate: A perspective gained from sedimentary organic geochemistry and particle size analysis. Polar Science. 2015: 9, 359-367.

[2] Mahesh B.S., Warrier AK, Mohan, R., Tiwari, M, Roy, R., Asthana, R., and Ravindra, R.,. Response of Sandy Lake in Schirmacher Oasis, East Antarctica to the glacial-interglacial climate shift. Journal of Paleolimnology. 2017: 58 275-289

[3] Mahesh, B.S., Nair, A., Warrier, A.K., Avadhani, A., Mohan, R., and Tiwari, M. Palaeolimnological records of regime shifts from marine-to-lacustrine system in a coastal Antarctic lake in response to post-glacial isostatic uplift. Current Science Special Section: Asian Forum for Polar Sciences. 2018: 115 1679-1683

[4] Mahesh, B.S., Warrier, A.K., Mohan, R., and Tiwari, M. Impact of Antarctic climate during the Late Quaternary: Records from Zub Lake sedimentary archives from Schirmacher Hills, East Antarctica Palaeogeography, Palaeoclimatology, Palaeoecology. 2019: 514 398-406.

[5] Warrier, A.K., B.S. Mahesh, Rahul Mohan, Ravindra, R. Glacial–interglacial climatic variations at the Schirmacher Oasis, East Antarctica: The first report from environmental magnetism Palaeogeography Palaeoclimatology Palaeoecology. 2014, 412, 249-260.

[6] Warrier, A.K., Pednekar, H., Mahesh, B.S., Gazi, S., Sediment grain size and surface textural observations of quartz grains in late quaternary lacustrine sediments from Schirmacher Oasis, East Antarctica: Paleoenvironmental significance. Polar Science. 2015:10 89-100.

[7] Warrier, A.K., Mahesh., B.S., Mohan, R., 2017. Lake Sediment Studies in Schirmacher Oasis and Larsemann Hills, East Antarctica – An Indian Perspective Proceedings of the Indian National Science Academy DOI:10.16943/ptinsa/2017/48950.

[8] Gore, D.B. Last glaciation of Vestfold Hills: extension of the East Antarctic ice sheet or lateral expansion of Sorsdal Glacier. Polar Records. 1997: 33 (184), pp.5-12.

[9] Verleyen, E., Hodgson, D.A., Sabbe, K., Vanhoutte, K., Vyverman, W. Coastal oceanographic conditions in the Prydz Bay region (East Antarctica) during the Holocene recorded in an isolation basin. Holocene. 2004a: 14, 246-257.

[10] Hodgson et al. 2005; Hodgson, D.A., Verleyen, E., Sabbe, K., Squier, A.H., Keely, B.J., Leng, M.J., Saunders, K.M., Vyverman, W. Late Quaternary climate-driven environmental change in the Larsemann Hills, East Antarctica, multi-proxy evidence from a lake sediment core. Quaternary Research. 2005: 64, 83-99.

[11] Gibson J.A.E., Burton H.R. and Gallagher J.B. Meromictic Antarctic lakes as indicators of local water balance: structural changes in Organic Lake, Vestfold Hills 1978-1994. In: Quilty P.G. (ed.), Climate Succession and Glacial History of the Southern Hemisphere over the Past Five Million Years. ANARE Research Notes 94, Australian Antarctic Division. 1995: 39 pp.

[12] Pickard, J. Antarctic Oasis. Terrestrial environments and history of the Vestfold Hills. Academic Press, Sydney Orlando San Diego. 1986: 367 pp

[13] Ravindra, R. Geomorphology of Schirmacher Oasis, East Antarctica.

Proceedings, Symposium on snow, ice and glacier Geological Survey of India. 2001: Sp Pub 53 379-390

[14] Ravindra, R, Chaturvedi, A., and Beg, M.J. Melt-water Lakes of Schirmacher oasis-Their genetic aspects and classification. Pub in Advances in marine and Antarctic Sciences, New Delhi, Ed D.B. Sahoo and P.C, Pandey. 2002: pp301-313.

[15] Sunil, P.S., Reddy, C.D., Ponraj, M., Dhar, A., and Jaypaul, D. GPS determination of the velocity and strain-rate fields on Schirmacher Glacier, central Dronning Maud Land, Antarctica. Journal of Glaciology. 2007: 53, 558-564.

[16] Fairbanks, R.G. A 17,000 year glacial euststic sea level record: influence of glacial melting rates on the Younger Dryas event and deep ocean circulation. Nature. 1989: 342, 637-641

[17] Krause, WE, Krbetschek, M.R., and Stolz, W. Dating of Quaternary lake sediments from the Schirmacher Oasis (East Antarctica) by Infra-red stimulated luminescence (IRSL) detected at the wavelength of 560 NM. Quaternary Science Reviews. (Quat. Geochronology). 1997: 16, pp.387-392.

[18] Hodgson, D.A., Noon, P.E., Vyverman, W., Bryant, C.L., Gore, D.B., Appleby, P., Gilmour, M., Verleyen, E., Sabbe, K., Jones, V.J., Ellis-Evans, J.C., Wood, P.B. Were the Larsemann Hills ice-free through the Last Glacial Maximum? Antarctic Science. 2001: 13, 440-454.

[19] Achyuthan, H., Asthana, Ravindra, R., and Eastoe, C. Radiocarbon dates and sedimentation within the Schirmacher Oasis, East Antarctica. Abstract SCAR's open science congress, St. Petersburg, Russia. 2008.

[20] Shrivastava, P.K., Roy, S.K., Beg, J., Asthana, R., Govil, P., and Verma, K. Transition in Late Quaternary Paleoclimate in Schirmacher Region, East Antarctica as Revealed from Lake Sediments. Journal of Geological Society of India. 2018. 91(6):651-663.

[21] Sharma, C., Chauhan, M.S., and Sinha, R. Studies on Holocene climatic changes from Priyadarshini Lake sediments, East Antarctica: the palynological evidence Journal of Geological Society o India. 2007: 69, 92-96.

[22] Bera, S.K. Late Holocene palaeo-winds and climatic changes in Eastern Antarctica as indicated by long distance transported pollen-spores and local microbiota in polar lake core sediments. Current Science. 2004:, v. 86 (11), pp.1485-1488

[23] Phartiyal, B. Holocene paleoclimatic variation in the Schirmacher Oasis, East Antarctica: A mineral magnetic approach. Polar Science. 2014: 8, 357-369

[24] Phartiyal, B., Sharma, A., Bera, S.K. Glacial lakes and geomorphological evolution of Schirmacher Oasis, East Antarctica, during late Quaternary. Quaternary International. 2011: 235:128-136

[25] Gillieson, D., Burgess, J., Spate, A., and Cochrane, A. ANARE Research Notes 74. An Atlas of the lakes of the Larsemann Hills, Princess Elizabeth Land, Antarctica. Australian Antarctic Division. 1990.

[26] Verleyen. E., Hodgson, D.A., Sabbe, K., and Vyverman, W. Late Quaternary deglaciation and climate history of the Larsemann Hills (East Antarctica). Journal of Quaternary Science. 2004: 19:361-375.

[27] Verleyen, E., Hodgson, D.A., Sabbe, K., Cremer, H., Emslie, S.D., Gibson, J., Hall, B., Imura, S., Kudoh, S., Marshall, G.J., McMinn, A., Melles,

M., Newman, L., Roberts, D., Roberts, S.J., Singh, S.M., Sterken, M., Tavernier, I., Verkulich, S., Van de Vyver. E., Nieuwenhuyze, W.V., Wagner, B., and Vyverman, W. Postglacial regional climate variability along the East Antarctic coastal margin—evidence from shallow marine and coastal terrestrial records. Earth Science Review. 2011: 104:199-212.

[28] Sabbe, K., Verleyen, E., Hodgson, D.A., Vanhoutte, K., Vyverman, W. Benthic diatom flora of freshwater and saline lakes in the Larsemann Hills and Rauer Islands (E-Antarctica). Antarctic Science. 2003: 15, 227-248.

[29] Sabbe, K., Hodgson, D.A, Verleyen, E., Taton, A., Wilmotte, A., Vanhoutte, K., Vyverman, W. Salinity, depth and the structure and composition of microbial mats in continental Antarctic lakes. Freshwater Biology. 2004: 49, 296-319

[30] Hodgson, D.A., Vyverman, W., Verleyen, E., Sabbe, K., Leavitt, P.R., Taton, A., Squier, A.H., and Keely, B.J. Environmental factors influencing the pigment composition of in situ benthic microbial communities in east Antarctic lakes. Aquatic Microbiology and Ecology. 2004: 37, 247-263.

[31] Verleyen, E., Hodgson, D.A., Sabbe, K., Cremer, H., Emslie, S.D., Gibson, J., Hall, B., Imura, S., Kudoh, S., Marshall, G.J., McMinn, A., Melles, M., Newman, L., Roberts, D., Roberts, S.J., Singh, S.M., Sterken, M., Tavernier, I., Verkulich, S., Van de Vyver, E., Nieuwenhuyze, W.V., Wagner, B. and Vyverman, W. Post-glacial regional climate variability along the East Antarctic coastal margin - Evidence from shallow marine and coastal terrestrial records. Earth Science Reviews. 2011: 104, 199-212.

[32] Hodgson, D.A., Noon, P.E., Vyverman, W., Bryant, C.L., Gore, D.B., Appleby, P., Gilmour, M., Verleyen, E., Sabbe, K., Jones, V.J., Ellis-Evans, J.C., and Wood, P.B. Were the Larsemann Hills ice-free through the Last Glacial Maximum? Antarctic Science. 2001: 13, 440-454.

[33] Hodgson, D.A., Verleyen, E., Squier, A.H., Sabbe, K., Keely, B.J., Saunders, K.M., Vyverman, W. Interglacial environments of coastal East Antarctica: Comparison of MIS 1 (Holocene) and MIS 5e (last interglacial) lake-sediment records. Quaternary Science Reviews. 2006: 25, 179-197.

[34] Hodgson, D.A., Verleyen, E., Vyverman, W., Sabbe, K., Leng, M.J., Pickering, M., Keely, B.J., 2009b. A geological constraint on relative sea level in Marine Isotope Stage 3 in the Larsemann Hills, Lambert Glacier region, East Antarctica (31 366-33 228 cal yr BP). Quaternary Science Reviews. 2009b: 28, 2689-2696.

[35] Verleyen, E., Hodgson, D.A., Vyverman, W., Roberts, D., Mcminn, A.,Vanhoutte, K. and Sabbe, K. Modelling diatom responses to climate induced fluctuations in the moisture balance in continental Antarctic lakes. Journal of Paleolimnology. 2003: 30, 195-215.

[36] Hodgson, D.A., Verleyen, E., Sabbe, K., Squier, A.H., Keely, B.J., Leng, M.J., Saunders, K.M., Vyverman, W., 2005. Late Quaternary climate-driven environmental change in the Larsemann Hills, East Antarctica, multi-proxy evidence from a lake sediment core. Quaternary Research. 2005: 64, 83-99.

[37] Mazumder, A., Govil, P., Sharma, S,, Ravindra, R., Khare, N., and Chaturvedi, S.K. A testimony of detachment of an inland lake from marine influence during the mid-Holocene in the Vestfold Hills region, East Antarctica Limnological Reviews. 2013a: 13209-214

[38] Mazumder, A., and Govil, P. Signature of warmer Late Holocene around Vestfold Hills, East Antarctica Canadian Journal of Basic Applications Science. 2013: 1 33-43.

[39] Govil, P., Asthana, R., Mazumder, A., and Ravindra, R. Grain size distribution and its influence on biological productivity during Holocene in a fresh water lake in Larsemann Hills, Antarctica. National Academy of Science Letters. 2012: 35 115-119.

[40] Asthana, R., Shrivastava, P.K,. Beg, M.J., Swain, A.K., Dharwadkar, A., Roy, S.K., and Srivastava, H.B. Sedimentary processes in two different polar periglacial environments: Examples from Schirmacher Oasis and Larsemann Hills, East Antarctica Special Publications in Geological Society of London. 2013: 381 411-427.

Chapter 3

Kalman Filter Harmonic Bank for Vostok Ice Core Data Analysis and Climate Predictions

Migdat Hodzic and Ivan Kennedy

Abstract

The Vostok ice core data cover 420,000 years indicating the natural regularity of Earth's surface temperature and climate. Here, we consider four major cycles of similar duration, ranging from 86,000 to 128,000 years, comprising 15% of periods for the warming interglacials compared to some 85% of cooling periods. Globally, we are near the peak of a rapid warming period. We perform a detailed frequency analysis of temperature and CO_2 cycles, as a primary stage in building a logical Climate Prediction Engine (CPE), illustrated with specific harmonics. This analysis can be repeated for all harmonics and various cycle combinations. Our time correlation estimates the CO_2 time lag for temperature at 400–2300 years, depending on the cycle, longer on average than previously concluded. We also perform Fast-Fourier transform analysis, identifying a full harmonic spectrum for each cycle, plus an energy analysis to identify each harmonic amplitude – to achieve further prediction analysis using a Kalman filter harmonic bank. For Vostok data we can use combinations of different cycles compared to the most recent for learning and then the current ongoing cycle for testing. Assuming causal time regularity, more cycles can be employed in training, hence reducing the prediction error for the next cycle. This results in prediction of climate data with both naturally occurring as well as human forced CO_2 values. We perform this detailed time and frequency analysis as a basis for improving the quality of our climate prediction methodologies, with particular attention to testing alternative hypotheses of the possible causes of climate change. These include the effect on albedo of suspended dust and increasing water vapor with temperature in initiating interglacial warming, the effect of temperature and pH values of surface water on ambient level of CO_2 in the atmosphere and finding a larger latent heat capacity in the atmosphere required to sustain its circulatory motions, leading to friction and turbulent release of heat in boundary layer. All these potentials can be examined in an effective CPE.

Keywords: Vostok data, time and frequency analysis, Kalman filter harmonic bank, climate prediction engine, machine learning

1. Introduction

Extensive climatic data for the past four ice ages and earlier, the period in which *Homo sapiens* evolved, is available in various scientific reports analyzing ice cores, commencing from the mid-1950s. There are various sites on Antarctica and

Greenland where intensive ice core drilling has occurred since 1956, with several countries supporting more than a score of different drilling projects. Currently, intensive ice core drilling is being conducted in other areas as well, so an even larger data set is anticipated. In previous papers [1–4] the history and limitations of ice core drilling are described in detail. Our purpose in this paper is to employ Vostok ice core data for time and frequency related analyses to set a basis for improving prediction of climate variations. The data sets include derivations of relative temperature, carbon dioxide (CO_2), methane (CH_4), oxygen, dust and solar variation (insolation), during the past 420,000 years. Because isotopic fractionation of oxygen-18 and deuterium in snowfall is temperature dependent and a strong spatial correlation exists between mean annual temperature and mean isotopic ratios it is possible to derive ice-core climate records. References [5, 6] presented the first record for full glacial–interglacial period from an ice core drilled in the Russian Vostok station in Antarctica. However, it is important to establish criteria for controlling the quality of this new science that we claim could eventually lead to a Climate Prediction Engine (CPE), based on verified causes.

A 420,000 year record was constructed from the [5] study on a 3 km deep core of ice (**Figure 1**). Another source of similar ice core data is available from European EPICA drilling project [7] which lasted from 1998 until 2005. EPICA data are comparable with Vostok data. In this paper we focus on specific analysis related to only two of the Vostok data variables, namely relative temperature and CO_2 content. We also touch upon possible effect of the dust on various points along Vostok timeline. We will include other data at a later date. Data sets used are from [8] and, with corrections, [9]. The variation of atmospheric CO_2, temperature and dust are shown in **Figure 1** together with our definition of four cycles (C1, C2, C3, C4) formed from the maxima of the variables. These cycles could also be defined starting from the variable minima. Before presenting very detailed time and frequency analysis of specific subset of Vostok data, we make some general observations related to climate fluctuations of CO_2, temperature and dust.

Variations in global climate measured by temperature change automatically involve the thermal energy content and heat capacity of the atmosphere. These highly variable systems can be contrasted with the energy content of conservative physical systems like planetary orbits, where the principle of least action defines the trajectories favored; conservative systems can be reliably modeled as an interaction

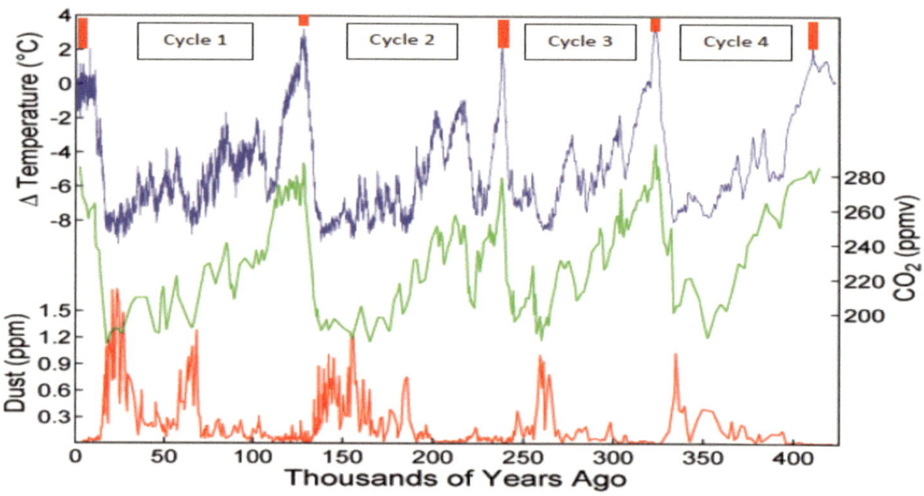

Figure 1.
Vostok data variation in global relative temperature and CO_2 content [8, 9].

between gravity and inertial forces, with the angular momentum and action held constant as a function of mass and the gravitational constant [10]. By contrast, the molecular states in the global ecosystem are dominated by transient energy flows, oscillating in a complex set of time scales as short as hourly to as long as geological epochs. Such variations in energy flows invite frequency analysis, characterized with Fourier transforms to provide underlying information.

A "snowball" Earth [11], more or less covered by ice and snow as the ice cores were fashioned clearly has diminished total energy in terms of molecular quantum states affecting molecular vibration, rotation, translation – as well as sustaining the more coherent circulatory motions in the atmosphere and the ocean, transferring thermal energy towards the poles. A warm Earth as at present has all these flows in higher quantum states, raising their entropy as we have quantified for atmospheric molecules elsewhere [12]. Variation in physical parameters such as temperature is obvious but complex in causes, with any longer-term trends overlying daily or seasonal trends as a function of latitude.

One state property that does not vary over time on Earth except geologically is atmospheric pressure, given that it is effectively based on the weight of the atmosphere, apart from variation for water, some 4% during the El Nino cycle [13]. The current IPCC consensus concludes that the most significant warming is from greenhouse gas content, such as that of CO_2, methane and nitrous oxide, but this is not necessarily the major controlling cause of increasing temperature, given how negotiable thermal energy is. Globally, we are conducting a large uncontrolled experiment to see if CO_2 is so important. Although originally thought that the CO_2 data from ice cores might be considered as proof of its causal role in global warming, but given that changes in CO_2 lag temperature changes its resultant release or absorption from solution in sea water is more likely the cause of correlations in the relationship.

An additional causative factor controlling the atmospheric pressure of CO_2 that seems to have been overlooked is varying acidity of waters, which can also respond to temperature. In warm periods with high oxygen levels, acidification by oxidation of reduced sulfur, nitrogen and carbon compounds is favored [14]. Alkaline conditions are favored anaerobically. The reduction of dimethyl sulfate, a major oxidant in sea water, can also lead to the evolution of reduced sulfur compounds, which can be converted to sulfuric acid in the atmosphere by ultraviolet oxidation. The alkaline pH of the ocean about 8.2 is effectively the same as that obtained in aerated water equilibrated with limestone ($CaCO_3$). Sources of alkaline carbonated salts in the ocean is basaltic volcanic rocks and soils, in contrast to more acidic granitic deposits that tend to acidify water.

We have shown using equilibrium theory that a lowering in surface ocean pH of 0.01 units can lead to an increased CO_2 pressure in the atmosphere of 8–10 ppm by volume. So, a change of 100 ppm could occur locally if the ocean surface or water on land was acidified by 0.10 unit, say from pH 8.20 to 8.10 as observed recently. Much larger local changes could occur, from a catastrophic event such as major volcanic eruption. The annual oscillation in CO_2 pressure at 3500 m altitude on Mona Loa, Hawaii could be a result of the change in the high surface temperature of the nearby ocean between winter and summer, rather than imbalance between photosynthetic assimilation and respiratory evolution of CO_2, often claimed in general climate models. The CO_2 pressure in Hawaii peaks after winter in May when the sea temperature is about 23°C and reaches its minimum in late October when the sea is about 27°C. A peculiarity of calcite is that its solubility declines with higher temperature [15], so its precipitation removing CO_2 from the atmosphere could partly or fully explain the oscillation. By contrast, at the sampling site at Cape Grim in Tasmania, where the annual sea temperature has a mean temperature of 15°C with variation between 11 and 19, a range which does not lead necessarily to precipitation of calcite, there is only slight evidence of an oscillation. While the burning of fossil fuels would be the main current cause of pH variation in the ocean, any other acidifying processes in the atmosphere or on land such as from agriculture, could also be controlling influences.

For all of the above reasons regarding the complexity of causes of climate change, we consider it would be beneficial if the ice core data could be subjected to careful Fourier frequency analysis, yielding detailed data regarding long term mechanisms of variation in climate, also pointing to an appropriate dynamic modeling of the underlying processes. In this chapter we focus specifically on CO_2 Vostok data for C1 as defined in **Figure 1**, as an example of our CPE approach, extending this to CO_2 for C2, C3 and C4, or their combinations. Other data such as temperature, methane, dust, and any other available ice core data such as EPICA, can be analyzed similarly.

2. Methodology

The eminent statistician Fisher [16] was an early exponent of testing statistical significance by harmonic analysis. Some preliminary spectral analysis has been conducted on Vostok ice core data set as reported in [17–19]. In this paper, a detailed spectral analysis in R and Excel was applied to Barnola et al. data set [8, 9]. In such analysis, time series are decomposed into underlying sine and cosine functions to establish the most important frequencies. Various texts on Fast Fourier Transforms (FFT) were also sources for frequency determination. These approaches allow construction of periodograms quantifying the contributions of the individual frequencies to the time series regression. The methodology developed for bioinformatics [20] has general application for time series and was employed in this study. In this paper we perform our own time and frequency analysis using R and Excel. Trudinger et al. [21, 22] have also applied Kalman filter analysis to ice core data. This allowed a more rigorous analysis of CO_2 variability for the Law Dome ice cores of Antarctica over the recent 1000 years than the usual deconvolution method. These authors pointed out that the Kalman filter allows better calculation of uncertainties in the deduced sources. The uncertainties correspond to the selected range of frequencies. They claimed [22] that it allows investigation of statistical properties that are directly related to physical properties.

Our aim is to apply Kalman filter based methodology to define the logic for a Climate Prediction Engine (**Figure 2**) based on 4 data cycles, **Figure 1**. Cycle C1 is the most recent period, the current interglacial warming period nearing its maximum but still incomplete. The overall number of published data points for both variables (relative temperature and CO_2) is 363. The individual cycles are determined by locating maximum absolute values for relative temperature and CO_2. Individual cycles differ in the number of data points slightly. Also, because of a lag observed between the maxima for temperature and CO_2 data, the number of data points differs slightly in two data sets in each cycle. This is also confirmed in our time correlation analysis below. **Table 1** summarizes the number of data points in each cycle. Note that each cycle is defined as top-to-top data values for both CO_2 and temperature, resulting in a different number of data points for CO_2 and temperature. However, the beginning of each interglacial where temperatures commences climbing might prove superior for some analyses, given that the controlling causes for reversal may be more consistent for these periods. Obviously, the current cycle is still evolving and new data may be added as required for further analysis. We can achieve that by appending the original Vostok data with additional current points, that might skew the previous natural data progression. Comparison of two

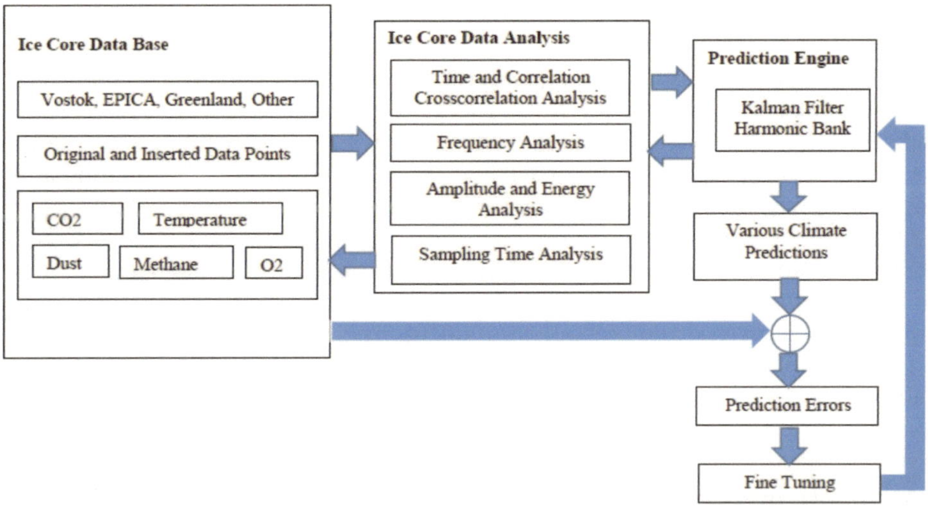

Figure 2.
Climate prediction engine (CPE) logic.

No of Data Points	Cycle C1	Cycle C2	Cycle C3	Cycle C4	Total
Temperature	74	136	99	54	363
CO_2	**80**	135	99	49	363

Boldfaced entries point to CO_2 and Cycle 1 related values.

Table 1.
Number of original Vostok data points for each cycle.

data sets including current data may be very useful in estimating global near and far future effects of factors like temperature and CO_2 content – an ultimate goal in this research.

Note that one such additional current data point may not make a large difference on the current C1, as far as harmonic analysis, which has long term time behavior built in, but the short term time effects may be more prominent. Our approach covers both short and long term effects and it can be used to perform sensitivity analysis using current climate data together with original Vostok data. The uneven data sampling times within each cycle) may pose some numerical issues as well for the analysis in the context of Kalman filter (KF) prediction methodology. There are various methods missing data problem and one of the simpler methods, yet effective one, is to enter missing data by some (linear or nonlinear) approximation method. We inject additional data using linear interpolation, also achieving Nyquist sampling requirements and in the process minimize the number of required harmonics used by the Kalman filter harmonic bank (KFHB). For C1 which originally had 80 data points for CO_2 we added additional data points for the total of 128 suitable for FFT but also corresponding to the length of 128,000 years so far in the cycle.

An aim is to gain more insight into Vostok data in time and frequency domains, and check the corresponding amplitude and energy content in order to reduce the number of significant harmonic components (**Figure 2**). Other authors used similar energy consideration but in our case, we propose the simple and effective mathematical model of a stochastic harmonic oscillator, based on which a KFHB can be built and used to further analyze Vostok data as well as predict near and far future data behavior. Note that besides C1 other cycles can be used to improve the precision by combining two or three cycles based on past data for C1, C2. C3. Amplitude and energy analysis performed here reduces a number of required individual Kalman Filter Harmonic Oscillators (KFHO) as individual blocks for the KFHB. It is important to note that using this approach rather than just combining amplitudes and harmonic components directly as described for example in [23], is to accommodate stochasticity of the data and the overall probabilistic nature of the prediction problem using KFs. This approach adds to the robustness of the method and provides better probabilistic accuracy both for short and long term periods, as well as allowing for prediction sensitivity analysis using various values, such as CO_2 levels as measured now and estimated for some future periods. Our methodology can be applied for both research as well as for policy making tools for the future climate related societal and technological decisions.

In this paper we perform time analysis of all cycles and C1 CO_2 frequency analysis (boldfaced in **Table 1**) to illustrate the KFHB approach (Section 7).

3. A climate prediction engine

The block diagram in **Figure 2** summarizes our CPE. It seeks the ability to predict various quantities using both past data (from one or more cycles) as well as the current data available This also allows to run various sensitivity analysis by using a variety of future scenarios as far as CO_2 and temperature, for example. The CPE can be applied to any ice core data, Vostok, EPICA or any other and for any variable, CO_2, temperature, methane, O_2, or dust.

The CPE consists of three major blocks. First one is a data base of various ice core data, both original, conditioned and inserted as required to make the data more uniformly distributed across the time span of each cycle. The second block is data analysis, in time and frequency domains, including various correlation measures. The third block is the prediction engine which consists of a set of oscillators (KFHO) that produce a KFHB (Section 8). Various predictions as well as sensitivity analysis are performed in this block. Finally prediction parameters are fine-tuned by original ice core data vs. the prediction errors. This is primarily done to fine tune KFHO gains. Section 8 contains all the mathematical details of KFHB.

4. Average sampling time analysis

First, we considered the entire set of Vostok data regarding time sampling. This will yield some initial indication of important issues in dealing with the data and how to perform further Vostok data filling to minimize the effects of non-uniform data distribution. As **Table 1** indicates the number of data points and the corresponding time periods for each cycle is quite different. This has to be taken into account when analysis is performed, as far as machine learning use of individual cycles for the benefit of estimating on going C1 data values. **Table 2** summarizes approximate duration of each cycle based on our definition of four cycles.

Duration in years	Cycle C1	Cycle C2	Cycle C3	Cycle C4
Temperature	127,726	115,156	86,462	96,782
CO_2	**128,399**	109,800	86,148	95,587

Boldfaced entries point to CO_2 and Cycle 1 related values.

Table 2.
Climatic cycles approximate duration in years.

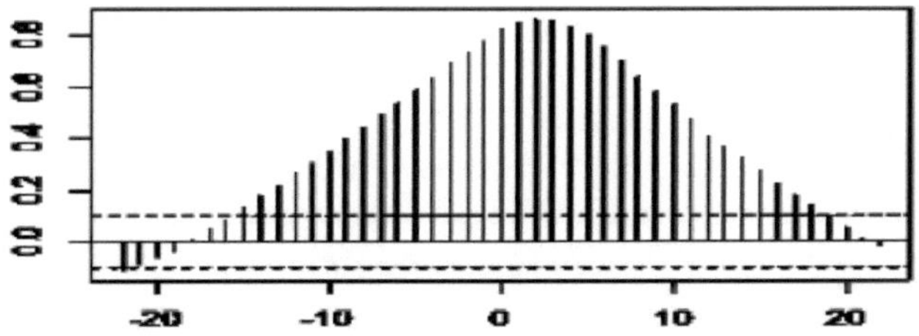

Lag	-2	-1	0	1	2	3	4	5	6
Value	0.734	0.778	0.821	0.847	0.863	0.859	0.835	0.802	0.752

Figure 3.
Short term temperature and CO_2 cross correlation.

Detailed visual inspection of maximum values in **Figures 2** and **3** indicates that CO_2 lags relative temperature in C4, C3, and C2. In C1, the data may be a little skewed due to the number of recent data points where maximum values are still not clearly identified in Vostok data (left-most CO_2 data, **Figure 1**), as C1 is still evolving. Better lag estimates can be obtained by cross correlation analysis as done in Section 4 below. Another feature is the varied time differences when the data is obtained from ice core readings. In some cases difference between two data samples are only 400–500 years, but in some as much as 5000 years. To start the analysis for each cycle we calculated average time sampling values as summarized in **Table 3**. In Section 5 this will be corrected to some extent by data filling and determining an ideal Nyquist and data sampling rate.

The differences in average sampling times obviously come from the number of data points collected and the duration of each individual cycle. To facilitate our general approach all other combination of available cycles data are also considered. The idea is enrich with the varied of all available data for prediction purposes and also training purposes of machine learning approach. For example **Table 4** indicates respectively average sampling times for C1 and C2 combined (C12), C2 and C3 combined (C23), as well as C2, C3 and C4 combined (C234), plus the total data set C1234. Note from **Table 4** that more cycles we add the more uniform average data sampling times become, until it becomes equivalent for both CO_2 and temperature. Further refinement of average sampling time is given in Section 5.2 bellow.

Average sampling time in years	Cycle C1	Cycle C2	Cycle C3	Cycle C4
Temperature	1703	800	865	1792
CO_2	**1605**	813	862	1911

Boldfaced entries point to CO_2 and Cycle 1 related values.

Table 3.
Average data sampling time in years for each cycle.

Average sampl. time, years	Cycles C12	Cycles C23	Cycles C234	Cycles C1234
Temperature	1087	827	1008	1149
CO_2	**1108**	834	1020	1149

Boldfaced entries point to CO_2 and combined Cycle C12.

Table 4.
Combined cycles C12, C23, C234 and C1234 mean data sampling times in years.

5. Time correlation analysis

As a starting point in time cross correlation analysis we examine correlation coefficients – single numbers that can be used as simple measure of cross correlation intensity between two variables. There are several coefficients named after their inventors such as Spearman, Pearson and Kendal [24] and they all indicate certain statistical properties that can relate two data series. **Table 5** summarizes standard cross coefficient between relative temperature and CO_2 for individual cycles as well as for the entire Vostok data set. The intensity of the cross correlation is quite high, on average more than 0.8 for the entire set. If we split the cycles into up and down sub cycles we obtain **Table 6** which indicates cross correlation coefficients for up and down cycle parts. Overall these coefficients indicate bigger spread between up and down sub cycles, and are very sensitive to where the break between up and down parts is chosen.

In general, one of the coefficients (up or down) is considerably larger than the overall single cycle coefficient. This might indicate that the usefulness of the individual up and down cross correlation analysis may be limited of the current C1 cycle. In the context of machine learning methodology this points to putting less emphasis on the cycles from longer in the past compared to the ongoing C1 cycle. To complete this analysis, the down-up period (boldfaced) ratios in **Table 6** indicate that the descending period is on average 6 to 8.6 times longer than the ascending period during interglacials, given that the cooling period is that much longer than the warming part. See also **Figure 9** for C1 CO_2.

Cross correlation coefficient	Cycle C1	Cycle C2	Cycle C3	Cycle C4	Entire cycle C1234
Temperature vs. CO_2	0.8389	0.8094	0.8069	0.8191	0.821

Table 5.
Cross correlation coefficient for individual and entire cycle.

Cycle 1	Down	Correlation	0.869	Down	Years	108,328
					Ratio	**7.05582**
	Up	Correlation	0.8262	Up	Years	15,353
	Total	Correlation	0.8389	Total	Years	123,681
Cycle 2	Down	Correlation	0.8008	Down	Years	97,087
					Ratio	**8.653802**
	Up	Correlation	0.9014	Up	Years	11,219
	Total	Correlation	0.8094	Total	Years	108,306
Cycle 3	Down	Correlation	0.8558	Down	Years	72,931
					Ratio	**5.961337**
	Up	Correlation	0.6503	Up	Years	12,234
	Total	Correlation	0.8069	Total	Years	85,165
Cycle 4	Down	Correlation	0.8519	Down	Years	83,533
					Ratio	**6.824034**
	Up	Correlation	0.827	Up	Years	12,241
	Total	Correlation	0.8191	Total	Years	95,774

Table 6.
Cross correlation coefficients for individual sub cycles.

Time correlation analysis produces a variety of useful information about periodicity and correlation strength among data samples of a given quantity. In particular autocorrelations produce the measure of self correlation of a data series, and the cross correlations indicate how two different data sets are correlated. We used standard programs such as R and Excel to generate those. One property of crosscorrelations R_{xy} is very useful in analysis of relative temperature vs. CO_2 content and that is the estimate of the time delay between the two. In general one needs to locate the maximum value point for R_{xy} and locate the corresponding argument, time lag in our case, between relative temperature and CO_2 content. **Figure 3** (with numerical values around zero lag) illustrates short term calculations for cross correlation between the two variables for the total C1234 cycle. We can read the value of the delay τ_{delay} between temperature and CO_2 as approximately equal to two lag units. Since the calculation is done for the entire data set, from **Table 4** we can read an average sampling time for C1234 as 1149 years, hence we can make a rough upper limit approximation of the size of the delay for the entire data set to be:

$$\tau_{delay} < 2 \text{ times } 1,149 \text{ years} \sim 2,300 \text{ years}$$

The calculation is approximate, primarily because the non-uniform distribution of the Vostok data. Some data is separated by only hundred years rather than thousands. This points to a need to make the data more uniform by inserting additional data. We address how to harmonize these data in Section 5. The total delay appears to be of order of 2000 or more years and not 100–200 years. That may be an important finding which can influence our thinking about the role of CO_2 increase caused by human actions. **Figure 4** indicates entire data set auto and cross correlation. It is clear that the data exhibits some periodicity. To determine average time delays between relative temperature and CO_2 for individual cycles we examine **Figures 5–8** which also have numerical values around zero lag. The first two diagrams in **Figure 5** are autocorrelations and they also indicate certain periodicity within the each cycle but obviously not as well as the entire data set. The third diagram shows cross correlation between two variables. The time delay can be read from cross correlation and for C1 it is less than one lag period but maybe more than zero lag, due to the non uniformity of data. We can estimate it as less than half of one period lag. From **Table 3** for both

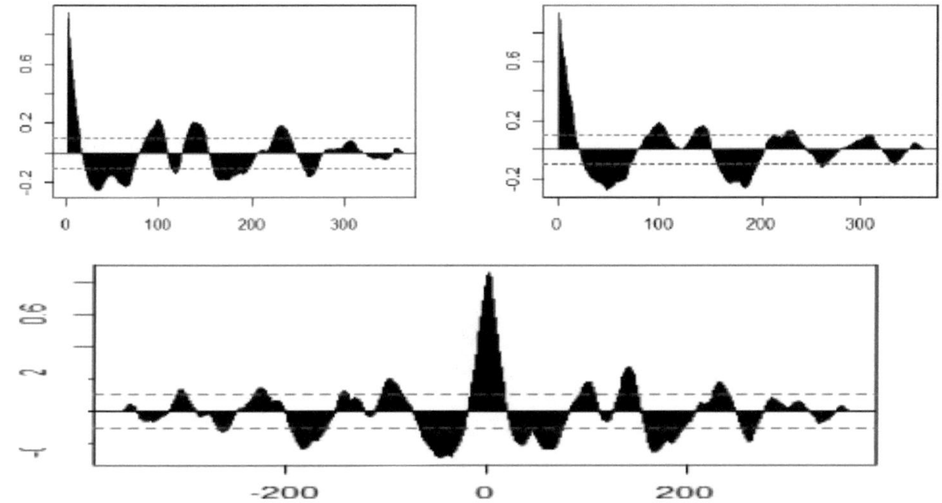

Figure 4.
Total long term temperature and CO_2 autocorrelations (left to right, above) and cross correlation (bellow) indicating inherent data periodicity.

relative temperature and CO_2 the average data sampling times are 1703 and 1605 years putting the absolute delay at around 800–850 years or less. Similar approximate estimates can be done for other cycles. Reading from C2 and C3 cross correlations, **Figures** 7 and **8** and **Table 3**, by the same consideration the delay is less than half the cycle, which translates into 400–435 years on average, or less. For C4 (**Figure 8** and **Table 3**) the delay appears to be of order of one cycle sampling time, a delay around 1800 years. To get a more precise approximations we would need more uniform data and finer resolution around the zero lag where the cross correlation is at its maximum. Note that on the individual cycle up and down parts this delay may differ from the average cycle level, also indicated by larger correlation coefficients spread in **Table 6**. We can identify no specific pattern in these coefficients regarding up and down sub cycles having larger or smaller coefficient. Overall, there is a significant CO_2 delay across total Vostok data C1234 compared to the relative temperature. For individual cycles a more detailed cross correlation analysis should be done, especially following data insertion. We made very rough estimates above.

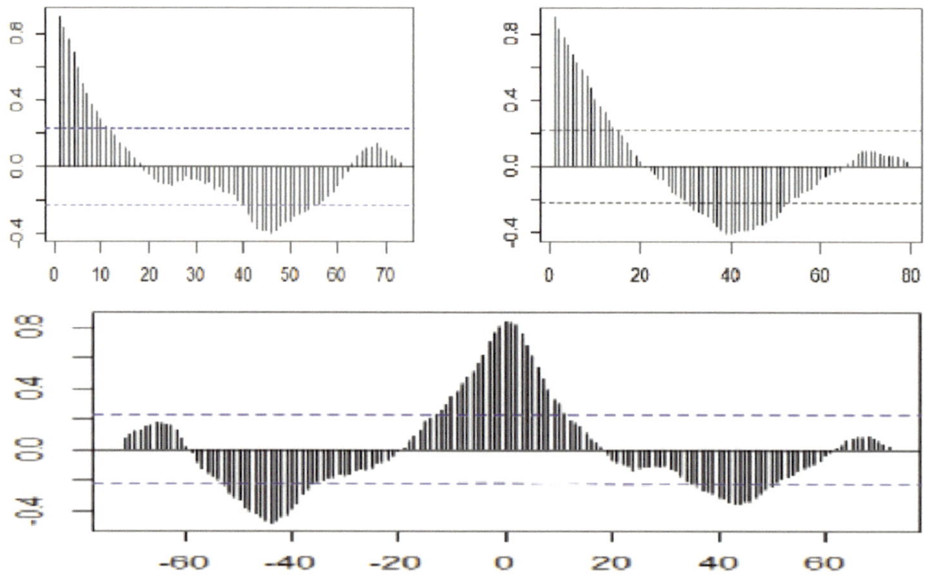

Lag	-2	-1	0	1	2		3	4	5	6
Value	0.766	0.805	0.839	0.834	0.818		0.758	0.682	0.614	0.546

Figure 5.
Cycle C1 temperature and CO_2 autocorrelations (left to right, above) and cross correlation (bellow) indicating some periodicity within the cycle C1 itself.

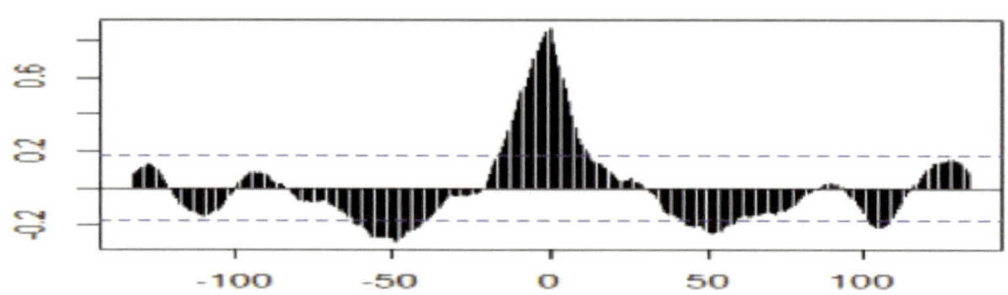

Lag	-3	-2	-1	0	1	2	3	4	5
Value	0.791	0.817	0.844	0.861	0.784	0.717	0.656	0.595	0.539

Figure 6.
Cycle C2 temperature and CO_2 cross correlation indicating some periodicity within the cycle C2 itself.

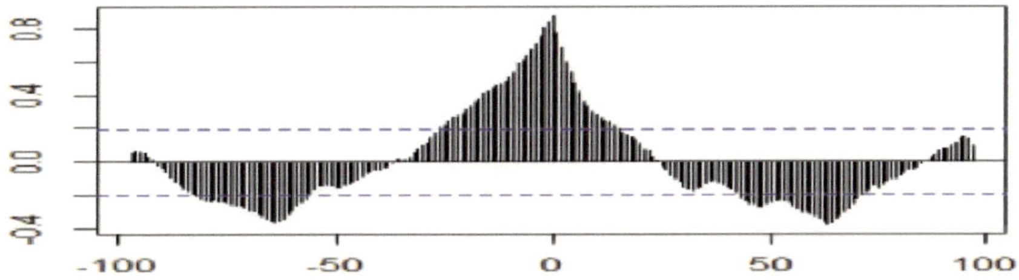

Lag	-6	-5	-4	-3	-2	-1	0	1	2
Value	0.639	0.675	0.709	0.754	0.805	0.840	0.876	0.774	0.687

Figure 7.
Cycle C3 temperature and CO_2 cross correlation indicating some periodicity within the cycle C3 itself.

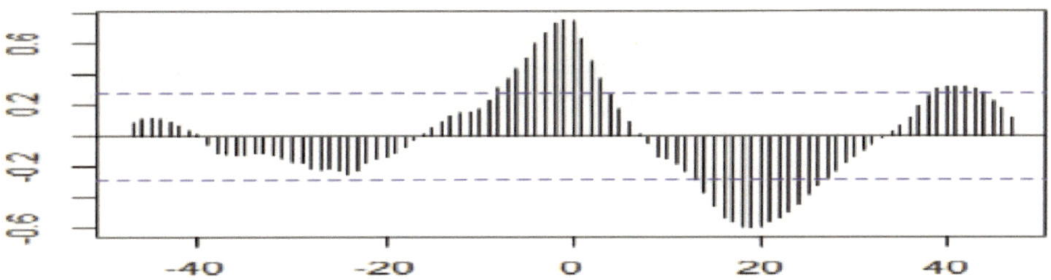

Lag	-6	-5	-4	-3	-2	-1	0	1	2
Value	0.439	0.510	0.601	0.671	0.728	0.754	0.745	0.628	0.485

Figure 8.
Cycle C4 temperature and CO_2 cross correlation indicating some periodicity within the cycle C4 itself.

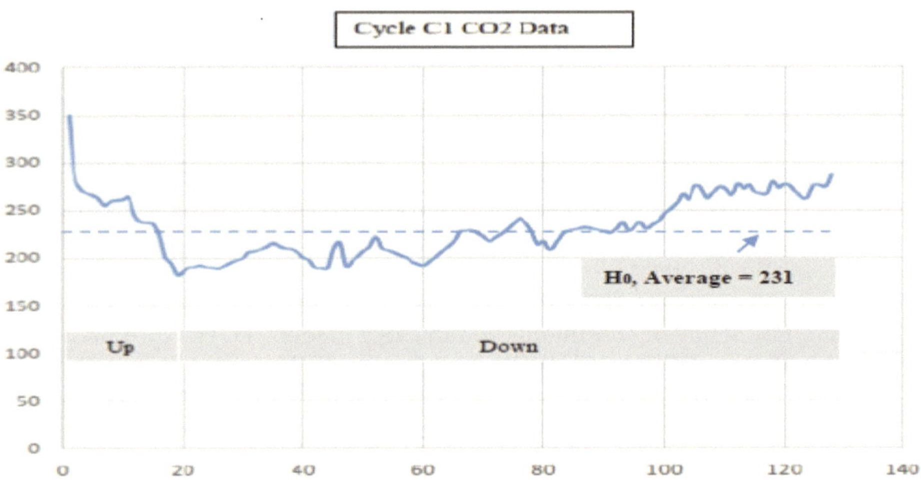

Figure 9.
Cycle C1 CO_2 content.

For practical reasons, the most important figure to keep in mind is the current cycle C1 delay which points to 800–850 years or less.

Note at the end of this Section that the way various cycles are defined (**Figure 1**) also affects the analysis. In a follow up work we aim at repeating this analysis by defining cycles from minim to minimum CO2 values. Choice of maximum or minimum values may assist in determining what triggers cycle reversals. We have early indications that the dust might have a significant role in this reversal.

6. Cycle C1 CO_2 frequency analysis

As an illustration of frequency analysis we examine Cycle C1 CO_2 data in details. Similar approach can be used for other data, such as temperature, methane, oxygen and dust. The first concern is to make the C1 CO_2 data more homogeneous hence the sampling time can be more precise and useful.

6.1 Data filling

The original Cycle 1 Vostok data contains a total of 80 CO_2 data points are very unevenly distributed in time. This poses issues with FFT or DFT harmonic analysis. Hence we corrected the situation somewhat by inserting certain number of additional "data points" using linear approximation between the original data points with the biggest time difference between the neighboring points, hence generating a total of 128 original and inserted data points. In particular, the very first point we added to the original Vostok data is the current value of CO_2 which we stated at 350 [25] for the year 1990, as an example. Current levels are estimated at over 400 ppm. Our CPE model can be used for a variety of sensitivity analysis by playing "what if" and processing the data through CPE.

Next data is the first original Vostok data, 2362 years in the past. This will correspond to 64 FFT harmonics described next. Note that the time period of C1 CO_2 data is 128,399 years (**Table 2**), time difference between the oldest C1 data and the current time, **Figure 9** above.

6.2 Sampling time

For sampling time, we proceeded as follows. As stated in Section 2 the cycle duration is measured between the maximum (CO_2 and temperature) points in each cycle. Hence, we have an average of a bit over 1000 years time sampling for Cycle 1, i.e. $T = \frac{128,399}{128-1}$ = 1011 years, which is used as the CO_2 data sampling time. The FFT analysis on 128 total points produces 64 harmonics with the highest frequency of $f_{max} = \frac{f}{2} = \frac{1}{2T} = \frac{1}{2022} 10^{-3} = 0.000495\, x\, 10^{-3}$ Hz. For the analysis we use $T = 1.011$ with the understanding it is in 1000s of years. **Table 7** summarizes Cycle 1 harmonic content ordered by the amplitude, from the largest to the smallest to identify each harmonic energy contribution. First harmonic H_1 is boldfaced. As shown in Section 7.1, for the discretization purposes, it is important to satisfy an additional sampling time requirement given in inequality (7) bellow which produces equivalent resonant frequency of continuous and discrete models, and in turn makes for more precise discrete KFHO and KFHB models as well. This increases required sampling frequency and reduces time sampling period bellow 1011 years, and we deal with it using energy analysis as described in **Table 8**. For other approaches to non-uniform data analysis see [26, 27] as well as general literature on non uniform and optimal Fourier analysis.

6.3 Amplitude and energy analysis

Table 7 shows that the average signal value (DC or H_0) is 231 and the rest of the energy content (reflected in a corresponding harmonic amplitude) of the harmonics follows the highest-to-lowest amplitude pattern H_1, H_2, H_4, H_3, H_7, H_{10}, H_5, and so on.

The cumulative column in **Table 7** indicates amplitude sum percentage of subsequent harmonics compared to the sum of all harmonic amplitudes. For example to reach 90% + of the total amplitude we need a total of 25 harmonics, including H_0.

H No	Frequency f	Period 1/f	Phase	Amplitude	Cumulative %
0	0	N/A	0	231.0068	58.83917
1	**0.007727498**	**129.408**	**0.754987**	**36.27814**	**68.07948**
2	0.015454995	64.704	0.555383	15.87177	72.12214
4	0.03090999	32.352	−1.11877	9.409128	74.51872
3	0.023182493	43.136	0.478204	6.815518	76.25468
7	0.054092483	18.48686	1.435168	5.576594	77.67508
10	0.077274975	12.9408	0.316939	5.098453	78.97369
5	0.038637488	25.8816	−1.14182	4.847574	80.20841
12	0.09272997	10.784	−0.20146	4.780349	81.426
6	0.046364985	21.568	−2.02131	3.49376	82.31588
17	0.131367458	7.612235	0.471006	3.462987	83.19793
14	0.108184965	9.243429	−0.58073	3.138693	83.99738
15	0.115912463	8.6272	0.489932	2.959198	84.75111
26	0.200914936	4.977231	−0.08324	2.30261	85.3376
25	0.193187438	5.17632	0.527865	2.031371	85.85501
38	0.293644906	3.405474	−0.48454	2.018039	86.36902
32	0.247279921	4.044	−0.171	1.841643	86.8381
11	0.085002473	11.76436	0.041037	1.827369	87.30354
34	0.262734916	3.806118	0.371439	1.794662	87.76066
37	**0.285917409**	**3.497514**	**0.402143**	**1.71814**	**88.19828**
60	0.463649852	2.1568	0.050809	1.694368	88.62985
21	0.162277448	6.162286	−1.80194	1.617602	89.04186
23	0.177732443	5.626435	0.231643	1.60276	89.4501
28	0.216369931	4.621714	0.236801	1.570401	89.85009
27	**0.208642433**	**4.792889**	**−0.21626**	**1.560695**	**90.24761**

Boldfaced entries correspond to the first harmonic H_1.

Table 7.
Summary of cycle 1 CO_2 harmonic analysis.

Highest frequency	Maximum frequency	Cumulative amplitude %	Total H's	Sampling Ts <	Chosen Ts
15	0.293645	86.369	15	1.083996	1.011
15	0.293645	88.1983	19	1.083996	1.011
20	0.46365	90.2476	24	0.686531	**0.650**
63	0.486832	96.6477	44	0.653839	**0.650**

Boldfaced entries point to the lowest limit on Sampling Time Ts.

Table 8.
Cycle 1 CO_2 amplitude and sampling time.

The last (boldfaced) harmonic in this list is H_{27}. For a bit smaller cumulative amplitude, say 88%+, we only need 19 harmonics, with H_{37} as the last one. **Table 8** further summarizes energy analysis. For example, a chosen sampling time of 1011 satisfies the requirement in inequality (7) bellow all the way to cumulative energy

MAPE	Harmonics						
	1	1 + 2	1 + 2 + 4	1 + 2 + 4 + 3	1 + 2 + 4 + 3 + 7	1 + 2 + 4 + 3 + 7 + 10	1 + 2 + 4 + 3 + 7 + 10 + 5
y	0.058	0.043	0.038	0.036	0.034	0.033	0.031
x Corrected	0.058	0.043	0.037	0.036	0.034	0.033	0.030
x Predicted	0.056	0.044	0.039	0.037	0.035	0.034	0.033

Table 9.
Kalman filter harmonic bank estimation errors.

of 88.1983%. To reach the energy level of 90.2476%, much lower sampling time would be required, i.e. 650 years instead of 1011. In order to satisfy this requirement we would need to almost double the total number of data points and use more harmonics. Fortunately. not all the amplitudes are at their maximums all the time, hence the above percentages are only very conservative lower bounds. In reality we can calculate the errors by MAPE (Maximum Absolute Percentage Error) and these will be much more realistic, actually of order of 3–5% off of 100%, per **Table 9**. Also, energy of these first seven harmonics is of order of 99% + due to its calculation based on amplitude squares. This analysis gives us an idea of various issues at hand related to non-uniformity of Vostok data as well as number of harmonics to be used based on energy analysis. Our aim is to show that properly designed KFHB based CPE can deal with these issues in a very effective way. In Section 7 we illustrate various points raised here using first 7 harmonics, H_1, H_2, H_4, H_3, H_7, H_{10}, H_5.

7. Kalman filter harmonic bank

In an illustrative example in this paper we focus on C1 CO_2 harmonic analysis and the corresponding KFHO for the strongest, amplitude wise, first harmonic H_1 in **Table 7**. The same analysis can be repeated for any harmonic as we elaborate in Sections bellow. We proceed with the development of KFHB which models a series of harmonic components obtained by the energy analysis, which approximates to a reasonable degree (say 90% + or higher cumulative amplitude percentage) the original Vostok data. First step is to define a general Markov model [28] for a generic harmonic oscillator in discrete time described in the following Section. Other authors also used the Kalman filter approach to analyze ice core data, but with a different emphasis, [21, 22] as mentioned in Introduction.

7.1 Discrete time harmonic oscillator

To start we introduce a continuous model of a Harmonic Oscillator described in [29, 30]:

$$\begin{pmatrix} \dot{x}_1 \\ \dot{x}_2 \end{pmatrix} = \begin{pmatrix} 0 & 1 \\ -\omega_0^2 & 0 \end{pmatrix} \begin{pmatrix} x_1 \\ x_2 \end{pmatrix} \tag{1}$$

where ω_0 is radial frequency, $\omega_0 = 2\pi f_0$, with f_0 frequency in Hz. The solution to the above state equation is sine or cosine function depending on the initial conditions. The state variables x_1 and x_2 are "position" and "velocity" of whatever variable we are dealing with, such as CO_2 content and its rate of change. The first level of continuous-to-discrete approximation is based on one point derivative approximation and sampling time T, whereas we obtain:

$$\begin{pmatrix} x_1(t+1) \\ x_2(t+2) \end{pmatrix} = \begin{pmatrix} 1 & T \\ -\omega_0^2 T & 1 \end{pmatrix} \begin{pmatrix} x_1(t) \\ x_2(t) \end{pmatrix} \tag{2}$$

with "$t+1$" standing for $(t+1)T$, and similarly $(t+2)T$ for "$(t+2)$", where T is dropped for simplicity. To the above discretized time model we can also add model uncertainty via additional stochastic zero mean Gaussian white inputs r_1 and r_2, with certain variance values, V_1 and V_2 which can be fine-tuned. Hence we have:

$$\begin{pmatrix} x_1(t+1) \\ x_2(t+1) \end{pmatrix} = \begin{pmatrix} 1 & T \\ -\omega_0^2 T & 1 \end{pmatrix} \begin{pmatrix} x_1(t) \\ x_2(t) \end{pmatrix} + \begin{pmatrix} r_1(t) \\ r_2(t) \end{pmatrix} \tag{3}$$

The initial conditions are given as a transposed vector $(x_1(0)|x_2(0))$ with:

$$x_1(0) = A_0\, cos\,(\theta_0), x_2(0) = -A_0\, sin\,(\theta_0) \tag{4}$$

where A_0 and θ_0 are the amplitude and the phase of the harmonic ω. Better discrete approximation can be obtained by 2 point derivative approximation whereas we obtain:

$$x_1(t+1) = -a_0 x_1\,(\mathrm{t}) - x_1(t-1) \tag{5}$$

where parameter a_0 is calculated to match discrete and continuous resonant frequencies:

$$a_0 = -2\cos\,(\omega_0 T) \tag{6}$$

It is important to note that in this case one needs to choose sampling time $T = 1/f$ to satisfy:

$$f > \pi f_0, \omega_0 = 2\pi f_0 \tag{7}$$

which is higher than the standard Nyquist frequency, $f > 2f_0$. We show in the next Section an example of this. Eq. (5) produces cosine function with the proper initial conditions $x_1(0)$ in (4). The corresponding equation for $x_2(t+1)$ is equivalent to (5) with the initial condition $x_2(0)$ in (4) to produce sin function. Instead of proceeding with two state model (3) we can produce another two state model using (5):

$$\begin{pmatrix} x_1(t+1) \\ x_1(t) \end{pmatrix} = \begin{pmatrix} -a_0 & -1 \\ 1 & 0 \end{pmatrix} \begin{pmatrix} x_1(t) \\ x_1(t-1) \end{pmatrix} + \begin{pmatrix} w_1(t) \\ w_2(t) \end{pmatrix} \tag{8}$$

$$y_0(t) = x_1(t) + v_0(t) \tag{9}$$

where $w_1(t)$ and $w_2(t)$ are zero mean Gaussian white inputs with joint covariance symmetric matrix Q = $\begin{pmatrix} Q_{11} & Q_{12} \\ Q_{12} & Q_{22} \end{pmatrix}$, [2]. The measurement $y_0(t)$ of $x_1(t)$ with the measurement error $v_0(t)$ is simply defined in (9) and it corresponds to the harmonic ω_0. Measurement error $v_0(t)$ is zero mean stochastic process with variance R_0 assumed constant across all time samples, which is a reasonable assumption, unless there is a compelling reason to make it time varying. The model (8) and (9) above remains the same. Obviously equivalent model holds for $x_2(t+1)$ with the proper initial conditions. The model as given by (8) and (9) is our starting point for KFHO described next. The consideration holds for any harmonic ω.

7.2 Kalman filter harmonic oscillator

To facilitate the next step, we rewrite (8) as:

$$\begin{pmatrix} x_1(t+1) \\ x_2(t+1) \end{pmatrix} = \begin{pmatrix} -a_0 & -1 \\ 1 & 0 \end{pmatrix} \begin{pmatrix} x_1(t) \\ x_2(t) \end{pmatrix} + \begin{pmatrix} w_1(t) \\ w_2(t) \end{pmatrix} \tag{10}$$

where $x_2(t+1)$ is just an auxiliary notation for $x_1(t)$ and it is not $x_2(t)$ in (3). Then the standard KF equations in the above case produce [2]:

Prediction Step:

$$\hat{x}_1(t+1/t) = -a_0\hat{x}_1(t/t) - \hat{x}_2(t/t) \tag{11}$$

$$\hat{x}_2(t+1/t) = \hat{x}_1(t/t) \tag{12}$$

Correction Step:

$$\hat{x}_1(t+1/t+1) = \hat{x}_1(t+1/t) + K_{11}(t+1)\tilde{y}_0(t+1) \tag{13}$$

$$\hat{x}_2(t+1/t+1) = \hat{x}_2(t+1/t) + K_{21}(t+1)\tilde{y}_0(t+1) \tag{14}$$

In (13), (14) above, $\tilde{y}_0(t) = y_0(t) - \hat{x}_1(t/t-1)$ is Innovation Sequence and filter gains are:

$$K_{11}(t) = P_{11}(t/t-1)/[P_{11}(t/t-1) + R] \tag{15}$$

$$K_{21}(t) = P_{12}(t/t-1)/[P_{11}(t/t-1) + R] \tag{16}$$

The corresponding Prediction Step and Correction Step variances and covariances of the estimation error $\tilde{x}_1(t/t-1) = x_1(t) - \hat{x}_1(t/t-1)$, and $\tilde{x}_1(t/t) = x_1(t) - \hat{x}_1(t/t)$, and similarly for the state $x_2(t)$, are:

Prediction Step:

$$p_{11}(t+1/t) = a_0{}^2 p_{11}(t/t) + 2a_0 p_{12}(t/t) + p_{22}(t/t) + Q_{11}. \tag{17}$$

$$p_{12}(t+1/t) = -a_0 p_{11}(t/t) - p_{12}(t/t) + Q_{12} \tag{18}$$

$$p_{22}(t+1/t) = p_{11}(t/t) + Q_{22} \tag{19}$$

Correction Step:

$$p_{11}(t+1/t+1) = [1 - K_{11}(t)]\, p_{11}(t+1/t) \tag{20}$$

$$p_{12}(t+1/t+1) = [1 - K_{11}(t)]\, p_{12}(t+1/t) \tag{21}$$

$$p_{22}(t+1/t+1) = -K_{21}(t)\, p_{12}(t+1/t) + p_{22}(t+1/t) \tag{22}$$

The initial conditions for the above equations are:

$$p_{11}(1/0), p_{12}(1/0), \text{and } p_{22}(1/0) \tag{23}$$

and they are determined by the initial Kalman Filter design. One way to determine them is to use matrix Q values:

$$p_{11}(1/0) = Q_{11}, p_{12}(1/0) = Q_{12}, \text{and } p_{22}(1/0) = Q_{22} \tag{24}$$

Here the values of Q_{11} and Q_{22} are assumed to be of same the order because they represent uncertainty in modeling $x_1(t)$ and $x_2(t)$, and they are just one step apart values of the same state. Simple correlation analysis of $x_1(t)$ and $x_2(t)$ indicates that Q_{12} is of order of $-a_0 Q_{11}/4$. **Figure 10** below shows a block diagram of a single ω_0 KFHO. Here we have the total state vector corresponding to specific harmonic ω_0 as:

$$\hat{x}_0(t/t-1) = [\hat{x}_1(t/t-1), \hat{x}_2(t/t-1)]^{\mathrm{T}} \tag{25}$$

7.3 Kalman filter harmonic bank

Once we define single harmonic KF as in **Figure 10** we can proceed and construct a KFHB as an assemblage of a number of individual harmonic filters in parallel with the combine outputs to form the original signal (data). We assume a set of harmonics $\omega = \{\omega_1, \omega_2, \omega_3, \dots, \omega_N\}$ and for each of ω_i, $i = 1, 2, \dots, N$ we define a separate KFHO as described above. Note that harmonics are related to each other via:

$$\omega_i = i\,\omega_1, i = 1, 2, \dots, N \tag{26}$$

and the total signal output (such as Votok data) is the sum of individual harmonic $\omega_i = 2\pi f_i$ outputs:

$$y(t) = \sum_{1}^{N} y_i(t) \tag{27}$$

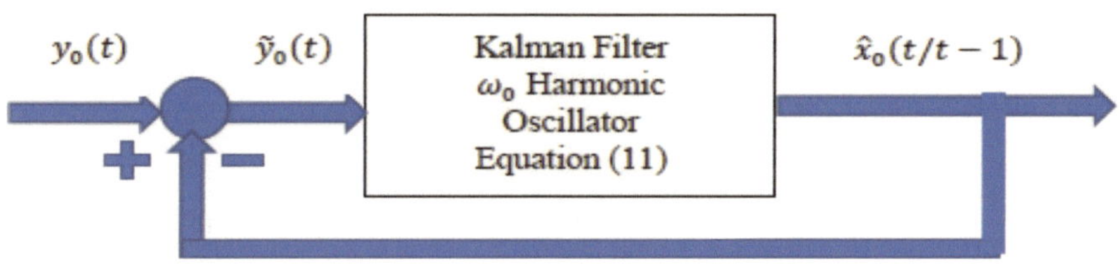

Figure 10.
Kalman filter harmonic oscillator.

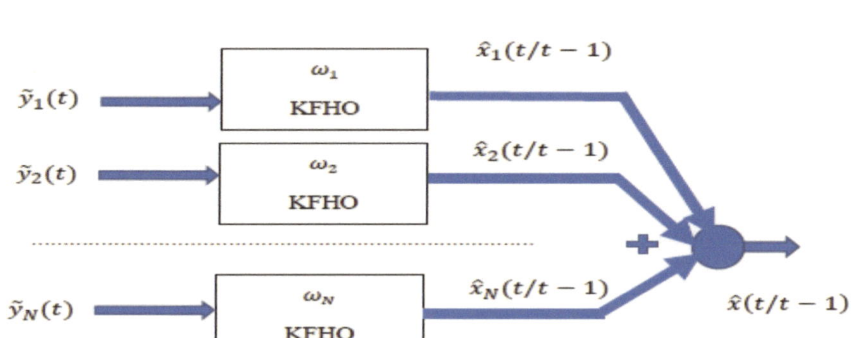

Figure 11.
Kalman filter harmonic bank.

Figure 11 indicates this arrangement with KFHO where we have:

$$\hat{x}(t/t-1) = \sum_{1}^{N} \hat{x}_i(t/t-1) \tag{28}$$

representing the total KFHB predicted state estimate. These can be used for short and long term prediction purposes for CO2, temperature or other variables of interest. We can similarly define a set of corrected state estimates which are more precise than predicted with the real data available.

$$\hat{x}(t/t) = \sum_{1}^{N} \hat{x}_i(t/t) \tag{29}$$

General idea here as compared to a simple sum of harmonic cosine signals (following inverse Fourier Transform) as in (27) with:

$$y(t) = \sum_{1}^{N} y_i(t) = \sum_{1}^{N} A_i \cos(\omega_i t + \theta_i) \tag{30}$$

is to accommodate stochasticity of the underlying Vostok measurement data as well as a simple linear structure of Kalman Filters Harmonic Oscillator, and its ability to make predictions for the signal future values in the probabilistic (more realistic) environment.

8. Vostok cycle C1 CO_2 Kalman filter harmonic oscillators

8.1 First harmonic H_1

The harmonic analysis of Cycle C1 Vostok data is presented in Section 5. In this Section we proceed and build a specific first harmonic H_1 KFHO, as an example of CPE set up, with the initial conditions:

$$\hat{x}_1(0/0) = A_1 \cos(\theta_1), \hat{x}_2(0/0) = -2\pi f_1 A_1 \sin(\theta_1) \tag{31}$$

Optimal	Constant Gain
$k_1(+)$	$k_2(+)$
N/A	N/A
0.96149	0.48075
0.93616	0.29119
0.93633	0.29926
0.93578	0.30137
0.9357	**0.30151**
0.9357	**0.30151**
0.9357	**0.30151**

Boldfaced entries indicate constant Kalman Filter gains.

Table 10.
Filter gains.

Q_{11}	100
Q_{22}	100
Q_{12}	49.759
R	25
a	−1.99759

Table 11.
Parameters.

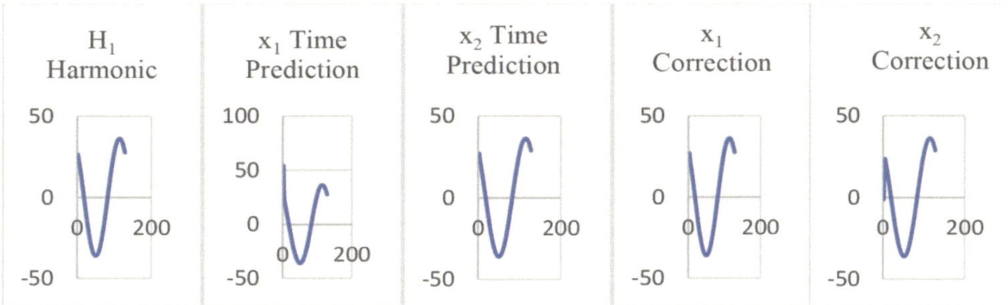

Figure 12.
Complete data, H_1 harmonic, predicted and corrected states.

$x_1(0)$	$x_2(0)$
Initial	Initial
231.0068	0
26.42065	−1.20707
13.48622	−0.81265
4.1098	1.643843
6.050971	−0.45685
0.754028	−1.87793
4.844518	−0.7715
2.016297	1.070199

Table 12.
Initial values.

were A_1 and θ_1 are first harmonic amplitude and phase. In **Table 10** we show H_1 KFHO gains from Eqs. (15), (16) for the first seven time samples. The boldfaced values indicate the moment (after 5 time samples) when the gains become constant. These constant gains can be used in the KFHO design for its simplicity. **Table 11** summarizes various filter parameters. The results for predicted and corrected state estimates $\hat{x}_1$ and $\hat{x}_2$ for 128 time samples are shown in **Figure 12**. The reference harmonic H_1 data is calculated using standard cosine function and it is the "measurement" as in (9) (**Table 12**).

8.2 Other harmonics (H_2, H_4, H_3, H_7, H_{10}, H_5)

The other harmonic calculations are done likewise, with specific amplitude and phase used and we will not give the obvious details here. The initial conditions are

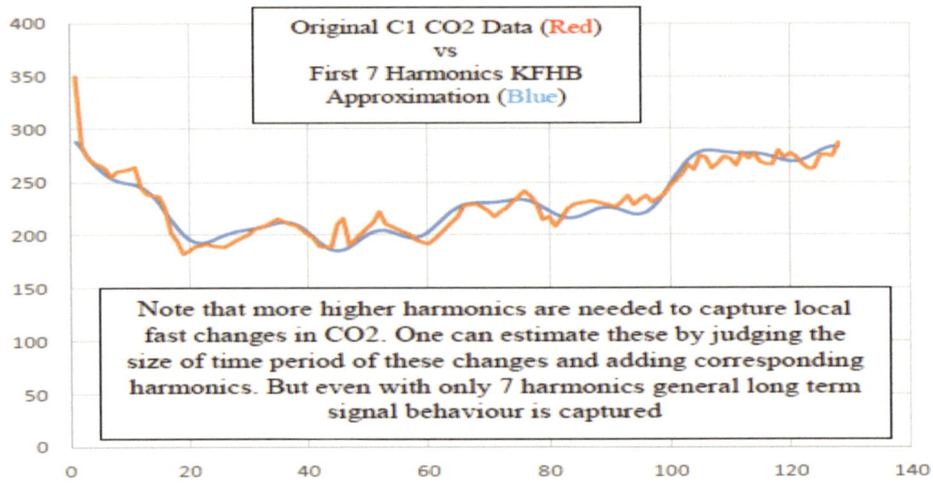

Figure 13.
Comparison of Vostok data vs. approximation with 7 harmonics.

calculated in the equivalent way as given in (31) using different values for amplitude, frequency and the phase angle, using **Table 7** harmonic values. Once calculated, initial conditions drive the KFHO for each harmonic, with the various harmonics KFHO parameters similar to the **Tables 11–12** above for H_1 and the other harmonic signals like in **Figure 12** but with different frequencies and phases. Next all of the predicted and corrected x_1 and x_2 estimates are combined per KFHB output in Eqs. (28) and (29). Resulting values are compared to the known original and inserted Vostok data and appropriate MAPE are generated to check the performance of the method. **Table 9** summarizes resulting MAPE for all 7 harmonics. Note that the results are based on constant filter gains in KFHB similar to the values in **Table 10** which are equivalent to the optimal time varying gains to at least 2nd decimal point. Parameters in **Table 11** were not optimized in any way and they can be tuned further based on errors obtained. This may be done so Q values would be adjusted per specific harmonic amplitude. We made a simple assumption that the standard model deviations and variances are of order of 100 for both states and all harmonics, and cross variance of order of $-a_0Q_{11}/2$, as noted earlier. Table also indicates what we expected, namely that the corrected errors are smaller than the predicted ones.

Note that the values for y are just specific harmonic amplitudes calculated based on maximum amplitude and phase angle, per Eq. (20). The prediction errors correspond to one sampling interval, in our case 1011 years. This is large enough for long terms calculations and prediction. If we chose to have 2 or more sampling intervals the prediction errors will obviously increase.

As far as corrected errors, they correspond to the situation where we have a specific CO_2 value to use for KFHB correction. For example if we chose to consider a current CO_2 value we can append it to the beginning of the Vostok data, and we actually did that by adding 350 PPM value for CO_2 'now' in front of the 'newest' Vostok data, more than 2000 years old. The CPE as we envisioned it allows for all sorts of scenarios, sensitivity analysis, 'what if' scenarios for all the variables, CO_2, temperature, methane, and so on, EPICA and other ice core measurements, both for short term (100+ years) as well as long term (1000 years+). To give a visual impression of Vostok data approximated by first seven harmonics, see **Figure 13** above. Approximation by 7 strongest amplitude harmonics produces very smoothed Vostok data (in blue) without capturing very abrupt changes on a smaller time scale, as in original Vostok data (in red). This can be improved by addition of more harmonics. The advantage which our CPE offers is its flexibility to treat both natural induced values of CO_2 and other variables as well as modern times human produced effects attributed to global warming.

9. Further considerations

Other cycle contributions can be examined as a base for KFHB based machine learning training and testing. This is important due to different duration of various cycles and the uncertainty about the duration of the ongoing Cycle 1. Then older cycles may carry less useful data for the predictions about the current cycle, especially in a view of current global warming effects due to human activities not present in the climate past. On the other hand the old data carries useful short and long term information which can be used judiciously in the KFHB fine tuning, in particular due to its inherent periodicity. Hence multicycle analysis can be very useful for machine learning implementation of our CPE based on KFHB idea. The minimal choice of harmonics will allow us to devise a reasonably simple machine learning algorithm for training, testing and prediction purposes based on KFHB. For example the longer data sets in C234 can be used as a training data set, as can shorter C23, in order to predict the completion of C1, calculating prediction error E1,234 (Error in predicting C1 given C2, C3 and C4 cycles) or E1,23 (predicting C1 given only C2 and C3 data). This applies to both temperature and CO_2. This can be repeated for other components in Vostok data set, such as methane, oxygen and insolation. Similarly for the European EPICA data set as well as set of cycles indicated in Milankovich theory [31]. Hence, once we predict C1 using C123 or C23, we obtain errors E1,234 and E1,23. Intuitively we can expect that E1,23 > E1,234, i.e. training based on larger data set ideally would produce smaller test and prediction errors. This has to be confirmed by the further analysis, in particular due to different lengths of the various cycles.

Besides our aim at producing an effective CPE methodology, the harmonic analysis spurred a variety of related thinking and ideas which we also summarize in this paper. Some further ones follow. Climate change on the time scales of the ice cores has been considered as consistent with the IPCC's hypotheses [11], focusing on permanent greenhouse gases, particularly CO_2, methane and nitrous oxide. This has included the role of increasing water vapor but viewed only as a secondary amplifying factor. The effect of temperature on water vapor pressure is shown by the Clausius-Clapeyron equation, dictating an exponential increase in vapor pressure as temperature rises. But the role of water vapor in modern global warming is only considered in GCMs as a derivative of primary warming by permanent greenhouse gases. This may be in error, as modern irrigation is now adding an extra 4% of water to land surfaces from 1960. It is possible to estimate water vapor content of atmospheres of different eras from temperature data. We have also hypothesized a positive forcing from irrigation water [32, 33] in addition to other primary sources of warming such as the Milankovich astronomical cycles. This may prove a more reliable means of correlation using the link already established between water vapor responsible for more than 80% of the air heating.

One feature of some of the ice core analyses is the irregular but rising increasing levels of dust as the planet became colder (**Figure 1**), possibly absorbing rather than scattering solar radiation. From frequency analysis, a marked impulse effect can be recognized, given that the peak of the dust samples in ice cores clearly coincides with the commencement of the interglacials, suggesting a role in initiating the warming process. Dust-climate feedbacks have recently been highlighted as having a role in the final stages of the glacial cooling process [34]; but we have surmised, the opposite that dust on the surface of snow and ice has decreased the albedo, capturing more radiant heat from the Sun. If true, a process in which a decreasing albedo effect from widespread dust in very cold, dry conditions (similar to the Antarctic now) initiates warming that is accentuated by increasing water vapor can be proposed. The rapid upward change of temperature in the interglacial would be directly consistent with the exponential increase in water vapor with temperature. In that case, the release of CO_2 could properly be seen as an effect of Henry's Law [35], rather than cause.

Consequently, the climate sensitivity for CO_2 may be overestimated. It would be of interest to know the scale of wind velocities in the glacials, since the inertial force of wind motion is opposed to gravity in the case of dust suspension and its carriage to high albedo surfaces. When dust particles are dry and disaggregated, current dust storms carrying particles above are most prominent. Furthermore, we have proposed [33, 36] that the atmosphere can store an order of magnitude more thermal energy because of the additional degree of freedom of motion in vertical motion. That is the motion involved in circulating air in anticyclones-cyclones. In a tropical cyclone, condensing water vapor is considered as providing energy to drive the cyclonic motion, using heat derived from the surface of the ocean. This follows from our finding [12] that configurationally entropy (the inverse of free energy) is a logarithmic function of the physical action, a scalar property related to angular momentum but including the dimensionless angular motion [10]. If confirmed this could mean that weather extremes such as very hot days with greater fire risk could be caused by collisions between anticyclones with extra thermal energy released as heat as the laminar flow of air becomes turbulent.

10. Conclusion

In this paper we have analyzed Vostok ice core data using (i) time correlations, (ii) harmonic analysis, as well as (iii) amplitude and energy consideration, and proposed a (iv) general prediction approach using KFHB methodology. In particular we focused on Cycle 1 of CO_2 data in frequency domain as a representative example. The general approach is to split Vostok data set into 4 smaller sets, as per climate periodicity indicated in the set. The outcome is a choice of set of high energy harmonics for all cycles and any of their combinations for designing CPE based on KFHB which is a linear combination of several individual KFHOs, for effective data prediction purposes. This can be incorporated into a practical machine learning methodology for training and testing, as well as data prediction using collected Vostok or other available climate data sets. We believe that our CPE based approach offers advantages in its simplicity and for short as well as long term prediction abilities via KFHB approach which can produce optimal results in the context of stochastic data set. Several issues remain to be solved, in particular (i) uneven cycle lengths as well as better approach to (ii) non uniformity of ice core data. We are addressing both in our ongoing work. Our analysis also seeks to find evidence regarding causes and results in climate science, an essential requirement for more certainty in weather and climate predictions. We consider that current GCMs have considerable uncertainties in such predictions.

DOI: http://dx.doi.org/10.5772/intechopen.94263

Author details

Migdat Hodzic[1]* and Ivan Kennedy[2]

1 American University in Bosnia and Herzegovina, Sarajevo, Bosnia and Herzegovina

2 Institute of Agriculture, University of Sydney NSW, Australia

*Address all correspondence to: migdathodzic@gmail.com

References

[1] Bentley CR, Koci R. Drilling to the beds of the Greenland and Antarctic ice sheets: A review, Annals Glaciology, 47, 2010.

[2] Langway CC Jr. The history of early polar ice cores. Journal Glacial Science, 55, 385–396, 2008.

[3] Fourteau K et al, Analytical constraints on layered gas trapping and smoothing of atmospheric variability in ice under low-accumulation conditions, Climates Past, 13,1815–1830, 2017.

[4] Alley RB, Reliability of ice-core science: historical insights. Journal Glaciology, 56,1095–1103, 2010.

[5] Petit JR et al, Climate and atmospheric history of the past 420,000 years from the Vostok ice core, Antarctica. Nature, 399, 429–436, 1999.

[6] Jouzel J et al, Vostok ice core: a continuous isotope temperature record over the last climatic cycle (160,000 years), Nature, 329,403–408, 1987

[7] https://en.wikipedia.org/wiki/, European Project for Ice Coring in Anta rctica, Accessed December 26, 2019.

[8] Barnola JM. Status of the atmospheric CO_2 reconstruction from ice cores analyses: Keynote perspective, Tellus, 51, 151–155, 1999.

[9] Barnola JM et al, Historical CO_2 record from the Vostok ice core. In Trends: A Compendium of Data on Global Change. Carbon Dioxide Information Analysis, Center, Oak Ridge National Laboratory, U.S. Department of Energy, Oak Ridge, Tenn., U.S.A. 2003.

[10] Kennedy IR, Action in Ecosystems: Biothermodynamics for Sustainability, Research Studies Press, Wiley, UK 2001.

[11] Pierrehumbert RT. Principles of Planetary Climate. Cambridge Press, UK 2010.

[12] Kennedy IR, Geering H, Rose M, Crossan. A simple method to estimate entropy and free energy of atmospheric gases from their action. Entropy, 21,451–464, 2019.

[13] Trenberth KE, Smith L. The mass of the atmosphere: A constraint on global analyses. Journal Climate, 18, 864–875, 2005.

[14] Kennedy IR. Acid Soil and Acid Rain. Research Studies Press, Wiley, UK 1992.

[15] Plath DC et al, The solubility of calcite – probably containing magnesium – in seawater, Marine Chemistry, 10, 9–29, 1980.

[16] Fisher RA. Tests of significance of harmonic analysis. Proceedings Royal Society A. 125, 54–59, 1929.

[17] Hodzic M and Kennedy IR, Time and Frequency Analysis of Vostok Ice Core Climate Data, PEN 7, 2, 907–923, 2019.

[18] Robert A. Michael at al, Temperature Impacts on Dengue Emergence in the United States: Ivestigating the Role of Seasonality and Climate Change, Epidemics 28, 2019.

[19] Yiou R, Baert E, Loutre MF. Spectral analysis of climate data. Surveys in Geophysics. 17, 619–663, 1996.

[20] Wichert S et al, Identifying periodically expressed transcripts in microarray time series data, Bioinformatics, 20:5–20, 2004.

[21] Trudinger CM et al, Kalman filter analysis of ice core data. Method

development and testing the statistics, Journal Geophysical Research,107:4422, 2002a.

[22] Trudinger CM et al, Kalman filter analysis of ice core data 1. Double deconvolution of CO_2 and delta^{13}C measurements. Journal Geophysical Research. 107:4423, 2002b.

[23] Fumi A et al, Fourier analysis for demand forecasting in a fashion company, INTECH International Journal Engineering Business Management, 2013.

[24] Boddy R and Smith G. Statistical Methods in Practice: For Scientists & Technologists, Wiley, US 2009.

[25] https://www.co2.earth/, Accessed December 26, 2019.

[26] Greengard L, Lee JY. Accelerating the Nonuniform Fast Fourier Transform, SIAM REVIEW, Society for Industrial and Applied Mathematics, 46, 443–454, 2004.

[27] Dutt A and Rokhlin V. Fast Fourier Transforms for Nonequispaced Data, Research Report YALEU/DCS/RR-980, 1993

[28] Anderson and Moore, Optimal Filtering, Prentice Hall, US 1986.

[29] Maarten van Walstijn, Discretization of the Harmonic Oscillator, PBASS, 2019.

[30] Cieslinski JL. On the exact discretization of the classical harmonic oscillator equation, Journal of Difference Equations and Applications, 17, 11, 2009.

[31] Meyers R, Sageman BB, Pagani M. (2008) Resolving Milankovitch: Consideration of signal & noise, American Journal Science, 308, 770–786, 2008.

[32] Kennedy IR, Hodzic M. Vortical action and entropy: A new proposal for the turbulent release of heat at the boundary layer. AMOS, Fremantle, 2020.

[33] Kennedy IR, Hodzic M. Designing a regional climate model to test the hypothesis that increasing anthropogenic use of water is a contributor to global warming. International Association Mathematical Geosciences Proceedings, Penn State, US, 2019.

[34] Shaffer G, Lambert F, Proceedings National Academy Science, 115,2026–2031, 2018.

[35] Bailey N et al, Henry's Law constant for CO_2 in aqueous sodium chloride solutions at 1 atm and sub-zero (Celsius) temperatures, Marine Chemistry, 207, 26–32, 2018.

[36] Kennedy IR, Computation of planetary atmospheres by action mechanics using temperature gradients consistent with the virial theorem. International Journal of Energy Environment, 9, 2308–1007, 2015.

Chapter 4

The Vegetation of the South Shetland Islands and the Climatic Change

Jair Putzke and Antonio Batista Pereira

Abstract

Antarctica allows at the same time to study the effects of change on the environment with minimal anthropic interference and in the least modified conditions in the world regarding biodiversity and its relations. At the same time, it allows assessing its effects on an ecosystem of few species and with a food web that directly links the oceans to terrestrial organisms. The South Shetland Islands are located further north within Antarctic Maritime and are therefore more vulnerable to climate change. Part of the studies already carried out with vegetation in this archipelago are discussed with a focus on the effects already generated and on predictions about future changes in the structure and plant diversity of Antarctica.

Keywords: vegetation, ecology, ice retreat, Antarctica, cryptogams

1. Introduction

The South Shetland Archipelago is located in the northern part of Antarctic Peninsula and is formed by 10 large islands (some reaching 100 km of lenght) and many smaller ones. The Maritime Antarctica, especially near the Antarctic Peninsula, have recorded the most significant temperature increases in the entire Southern Hemisphere, with 0.34° C per decade in the South Shetland Islands and between 1 and 1.4° C per decade (recorded since 1980) at the Rothera research station on the Antarctic Peninsula [1]. Data indicate that the marine water around the Antarctic Peninsula is up to 3° C warmer on average, contributing up to 50% of the ice melting already recorded [2].

The Antarctic Peninsula region had one of the most intense climatic warming trends over the last decades (increase of 0.56° C/decade in air temperature and 3°C in surface temperature since 1950) [3–5]. A statistically significant (at 3%) increasing trend in temperature was observed during the years 1944–1996, when the temperature increased by 1.6° C by analyzing the temperature in King George and in Deception Islands, from the South Shetland Archipelago. But in regions as the Admiralty Bay in King George Island the mean temperature was higher than 0.7° C, and the Wanda Glacier located there is retreating fast, having lost already 31% of its volume (compared to 1979) attributed to the regional warming [6].

The use of modeling proved that the ice river of the Thwaites Glacier that drains into the Amundsen Sea, in Western Antarctica, is already destabilized. The melting of these glaciers will raise the sea 1.2 meters on the planet, but the process will be very slow, probably hundreds of years [7].

Radar data proved that Pine Island Glacier retreated 31 km between 1992 and 2011, but has now reduced this speed. Until 2009 nothing was recorded about this retreat, when the melting and destabilization of glaciers suddenly began. The Larsen Glacier was one of the first to indicate a retreat (having persisted for 10,000 years), started to fall apart in 2002, collapsing in a period of 35 days and it is expected to disappear in 18 years [2].

In a reconstruction of changes in ice since the last glacial maximum, having studied at least 674 glacier data across the Antarctic Peninsula, it has been demonstrated as an environmental factor (the increase in the temperature of seawater rather than the atmosphere), was directly related to the retreat of glaciers [8]. The north–south gradient of increased retreat in the glaciers has a high correlation with ocean temperatures, since the water is cold in the Northwest and becomes progressively warmer at depths below 100 meters to the south. These waters of medium depth in the southernmost regions have been warming up since the 1990s, at the same time that the acceleration in the retreat of the glaciers began. And these waters are reaching lower and lower depths and affecting the emerged parts, as they heat up the platform. Almost all of the glaciers studied have declined since 1940.

The eastern region of the Antarctic continent, on the other hand, is slightly different from the region of the peninsula, since it is on dry land and has very thick ice. But because it is a more remote region, few scientists venture into the area and little data has been collected. However, data recently gathered from satellites and airplanes show another scenario. The Totten Glacier, for example, seems to be one of the most vulnerable, with the radar showing that there is a channel in the depths of it, which allows the entry of hot sea water that melts the ice and explains the loss of mass. This glacier can contribute to an increase of up to 3.5 meters in sea level [9].

Plant species on this continent are restricted to ice-free areas (except for microscopic algae that can grow directly on the ice) and are formations very threatened by climate change, as they do not support temperature changes very well. At the same time, plant communities are advancing in areas recently exposed by the retreat of ice and more favorable temperatures, resulting in the so-called "Antarctic greening". Analyzing five cores at three sites over 150 years, revealed increased biological activity over the past ca. 50 years, in response to climate change, suggesting that terrestrial ecosystems will alter rapidly under future warming, resulting in a greening similar to that registered to the Arctic [10].

It is important to note that a considerable carbon reservoir exists in cryobiont algae, which form extensive colonies directly on the ice. With the increase in temperature, it is expected that 62% of the blooms of small islands (like in the South Shetland archipelago) of low altitude will disappear [11].

There are at least 3 ways in which organisms can adapt to changes in the environment: 1- they can use the margins of physiological flexibility and then support changes. 2- can change the range of biological capacity which is highly dependent on the magnitude and rate of change. This ability is linked to the organism's reproductive capacity, but mutation rates, number of reproductive events and generation time are also linked. 3- they can migrate to have more favorable conditions. For Antarctic plants, the problems to be faced are greater to adapt, as they do not have an efficient disperser except the wind for lichen and moss spores, and must compensate locally for the differences to survive. And perhaps one of the big problems is getting nutrients. These are brought to the continent basically by animals, from their diet consisting of marine organisms [12]. A schematic of the flow of nutrients to the terrestrial environment can be seen in **Figure 1**.

Penguins are climate indicators and changes in their populations have been described over the past 50 years, mainly associated with changes in ice dynamics [13, 14]. *Pygoscelis adeliae* (Adélie penguin) is the most dependent on ice and the most

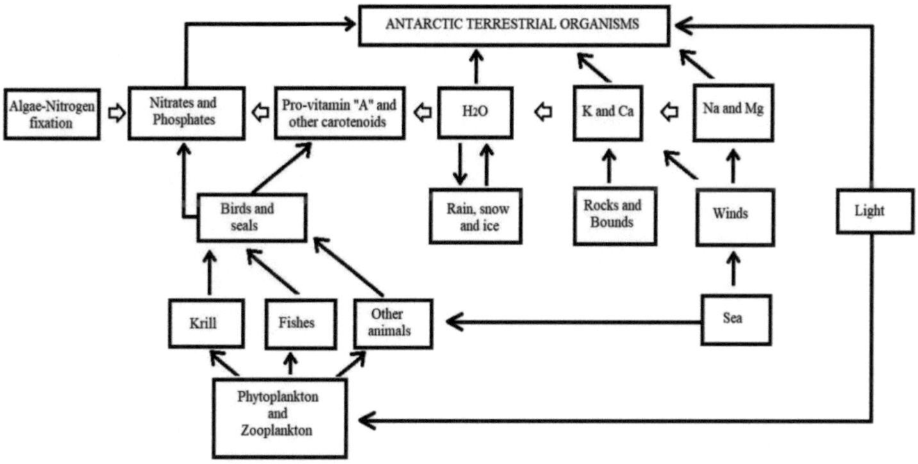

Figure 1.
Schematic view of the contribution of nutrients for the terrestrial ecosystems (adapted from [12]).

widely distributed species, occurring throughout the continent, as it is circumpolar. *P. antarctica* (the Antarctic Penguin) is found almost exclusively in the Antarctic Peninsula [15] and *P. papua* (the Papua Penguin) occurs in both the Antarctic peninsula and sub-Antarctic islands. The latter species has been showing its Antarctic populations expanding rapidly in the last 50 years, which is being associated with the increase in temperature in the region. Changes in ice dynamics have allowed the species to move further south, while *P. adeliae* and *P. antarctica* had a decrease in their populations mainly because availability of Krill, its main food. Adélie penguin populations are decreasing throughout the Antarctic Peninsula but they have remained stable on the east side of the continent, where the influences are still not so felt. In the past this species seems to have resisted climate change better. If it continues at this rate, it is estimated that populations can be reduced by 30% by 2060 and 60% by 2099 [16].

Fewer penguins means less availability of nutrients as they are one of the main sources of guano for the continent. Therefore, changes in plant communities that depend on this input may not happen. The melting of glaciers ends up exposing areas with rocks and sediments that will allow the installation of terrestrial vegetation, but nutrients must be available specially for the nitrophilic species.

Even cryobiont algae are found in ice, because it receives a spray of nutrients from penguins existing at least 5 km away [11]. The melting of the ice allows the melting water flows to carry these nutrients to the plants that grow on its banks, especially species like *Wanstorfia* spp., *Brachythecium* spp. and *Sanionia* spp. [17].

The Pinnipedia also have their contribution to the terrestrial environment, especially during periods when they are on land to rest. The deposition of feces and urine helps to nitrify the ground, but trampling can be harmful. A case described for the Signy Islands exemplifies this aspect, where the population of *Arctocephalus gazella* (fur seal) has greatly increased in recent years, completely destroying the existing vegetation in an area of about 80 hectares [18–21].

2. Plants with flower from Antarctica

Antarctica has only two native plants forming flowers: *Deschampsia antarctica* Desv. (Poaceae - **Figure 2**) and *Colobanthus quitensis* Kunth. (a Caryophyllaceae - **Figure 3**). There are already records of other Angiosperms occurring in the region, but these have been introduced by man, such as *Poa annua* L., and climate change can contribute to

Figure 2.
Deschampsia antarctica*, the Antarctic grass. Scale = 20 cm.*

Figure 3.
Colobanthus quitensis *among mosses and rock fragments.*

the occurrence of more plants in the region [22]. These two plants compete for space with all other species, but because they are larger and more complex, they need a large availability of nutrients and water. Therefore, they are usually found close to sources of nitrogen, such as in the vicinity of penguin rockeries or nests of other birds. Both have a chemical arsenal to survive the conditions of the Antarctic cold, especially a reasonable concentration of sugars in their cells: there is at least ten times more sugar in vacuoles than in sugarcane, foreseeing a potential use of this source in future. This accumulation of sugar is a protection against very cold periods in Antarctica [23].

Regarding the distribution of these phanerogams in the study area, it is possible to mention their occurrence in almost all the South Shetland Islands and areas of the Antarctic Peninsula free of ice. They can occur as small isolated tufts of a maximum of 15 cm, or forming fields of a few meters, but almost always associated with different Bryophyta and Marchantiophyta. Carpets even seem to stimulate grass development, but not its survival [24].

There are studies reporting the photoprotective effect of *Deschampsia antarctica* and *Colobanthus quitensis* extracts against UVB. The photoprotective properties have been attributed to several molecules, such as flavonoids and carotenoids, which absorb UV and act as antioxidants [25, 26].

It is possible that changes in temperature may interfere with the growth and development of populations of these species as has been shown experimentally [27, 28]. In the Argentine Islands, an increase of 25 times for *D. antarctica* and 5 times for *C. quitensis* was recorded in 30 years of observation [29]. Data collected in 2009 and historical data since the 1960s on the distribution of the two Antarctic vascular plants on Signy Island revealed that *D. antarctica* increased its coverage by 191% and the number of occurrence sites by 104%. *C. quitensis* increased its coverage by 208% and the number of occurrence sites by 35%. All due to the increase of 1.2° C in the air temperature and all the changes that this caused in the region [30].

Studying the formations of these phanerogams in the Fildes and Coppermine Peninsulas, in addition to locations in the Antarctic Peninsula in order to assess their responses to the increase in local temperatures, it was discovered that the populations of *D. antarctica* are expanding in the South Shetland Islands, but this expansion is not continuous in the Antarctic Peninsula, as the plants disappeared at 3 points, suggesting that there are other biotic and abiotic factors involved [31].

The fauna and flora associated with these plants is also very rich. There are bacteria, fungi and microscopic animals, many with a symbiotic or survival relationship with these plants. A high mortality of terrestrial microbial communities was detected along the South Shetland Islands. These communities are said to be dying from physiological problems and lack of nitrogen, in addition to changes in their microstructure, which seems to be associated with the rupture of the biogeochemical gradient of the microbial ecosystem. Caused by a strange but high abundance (explosion) of the associated fungi and the physical changes caused by them. All of these changes are related to the high temperatures recorded in the region. Some new diseases have been registered, especially for Antarctic grass, indicating that something is making possible the occurrence of these phytopathologies, but more studies are needed [32, 33].

There are also birds, of which at least the skuas (*Catharacta* spp.) and the gulls (*Larus dominicanus*) use these plants more frequently to make their nests. In a survey, scientists identified the seagull's preference for *Deschampsia antarctica* at Cierva Point in the Antarctic Peninsula [34]. More or less availability of this raw material can affect the reproduction of these birds.

These plants can be found in reproduction, but in general they are sterile. But higher average temperatures can contribute to increasing seed maturation, germination and seedling survival, although this has not yet been proven experimentally [35, 36].

3. Mosses and hepatics

Among the species that most stand out on more consolidated areas and even on Antarctic rocks, are mosses. The group that represents the bryophytes also has some liverworts occurring, but in this text, all will be commonly called mosses. There are, therefore, Marchantiophyta, popularly called hepatics, and the representatives of the genus

Marchantia are those that present the largest gametophyte (**Figure 4**), although small species of other genera sometimes take very large areas. Large populations have been found recently, such as the rare *Hygrolembidium isophyllum* in Harmony Point - Nelson

Figure 4.
Two Marchantiophyta, the thallose Marchantia berteroana *(above) and the leafy* Cephalozia *sp. (below).*

Island [37]. *Marchantia* is thallose, reproducing basically by direct fragmentation of the thallus or by specialized structures, the propagules, formed in receptacles such as in the figure (called conceptacles). But the group most represented in species in the area are the leafy liverworts (about 22 species). They even have a relationship with other organisms, as in the case of *Cephaloziella varians*, which is associated with a mycorrhizal fungus *Rhizoscyphus ericae* (ericoid symbiosis) throughout Antarctica [38–40].

Many species of liverworts are associated with dominant species in the plant community, and this reflects an interdependence. If the dominant species are threatened, by climate change, for example, their dependents will also be [41].

Bryophyta, or mosses themselves, have so far collected 113 species, within 55 genera and 17 families [18, 42]. The mosses present two main forms of growth: the pleurocarpic, where the moss stalk is prostrate, forming continuous carpets and in general covering more extensive areas if they are available (**Figures 5** and **6**);

Figure 5.
A moss carpet moved by wind being fixed by a scientist.

Figure 6.
Two large carpet of Sanionia uncinata *associated to* Warnstorfia sarmentosa *in the wettest areas.*

and the acrocarpic form, where the mosses grow upright, forming tufts or smaller cushions (**Figures 7–11**). The moss species with the highest occurrence and highest biomass in all ice-free spots is the pleurocarpic *Sanionia uncinata*, a carpet former with curved leaves, twisted like a scythe [18].

Antarctic moss fields can be very old and even deeper layers of growth can be alive even though they have been buried for over a thousand years by acrocarpic development. In 2014 research showed that the moss *Chorisodontium acyphyllum* remained alive after remaining frozen for more than 1500 years. A 1.4-meter-thick tuft was

Figure 7.
Tufts of the moss Syntrichia *sp. (red circles) growing among whale bones and a carpet of* Sanionia uncinata.

Figure 8.
Polytrichastrum alpinum*, one of the tallest moss (left) and* Pohlia cruda *(right).*

Figure 9.
Bartramia patens, *with sporophyte (left) and* Bryum palescens *(right).*

Figure 10.
Hennediella heimii *with sporophyte.*

Figure 11.
Chorisodontium acyphyllum, dull green and acrocarpic surrounded by a light green carpet of Sanionia uncinata *(pleurocarpic).*

sectioned every 20 cm (layer by layer) and placed to germinate under ideal conditions. In up to 8 weeks everyone started growing. The deepest layer was dated by radiocarbon and estimated between 1533 and 1697 years [43]. Acrocarpic and pleurocarpic mosses buried under more than 600 years by a glacier were re-exposed by the retreat of the ice and parts of the moss were able to activate again and grow normally "in vitro" [44].

The most available substrate in Antarctica is rock, but there are species that grow in soil. *Andreaea*, with four acrocarpic species occurring in Antarctica, for example, is exclusively saxicolous (name given to species that grow on rocks) [18].

Mosses are capable to colonize areas such as those closest to the sea and even receiving splashes of salt water waves, to the interior of the continent and including areas of recent exposition by ice retreat. The species *Muelleriella crassifolia* P. Dusén requires at least some contact with marine spray to develop, which is achieved in some coastal rocks. Eventually it can be found with other groups that have species with this preference, called halophytes, such as some of the lichens of the genus *Verrucaria* [45, 46]. The alterations in sea level will affect directly this community.

The rise in temperature in Antarctic regions has been accelerating the growth of mosses in particular since the late half of the 20th century. A study of 5000 years old population of *Polytrichum strictum* (in Lazarev Bay, on Alexander Island), dated by radiocarbon millimeter by millimeter, demonstrated that the population accumulated around 1.25 mm/year in the 19th and early 20th century and then increased its growth from 1955 until reaching 5 mm per year until the end of 1970, currently reducing growth to 3,5 mm/year. The authors also found that the associated amoeba population also increased considerably over the same period [47].

Studying a 1500 km gradient from Antarctic Maritime to the south of the Antarctic Peninsula (in the region of Lazarev Bay, Alexander Island) the accumulation in the banks of moss began to increase around the 1950's, reaching peaks in the Lazarev Bay in the 1970's (about 0.1 g of dry matter/cm^2/year) and Signy Island in the 1990s (0.06 g/DM cm^2/year); the most recent measurements indicate around 0.04 g of dry matter/cm^2. In continental Antarctica the growth of mosses is inversely proportional to the speed of the summer wind and proportional to the number of days above 0° C and the temperature of the summer [48].

Data collected in the Windmill Islands show evidence that the endemic moss *Schistidium antarctici* is likely to be more susceptible to climate change than the co-occurring and cosmopolitan species such as *Ceratodon purpureus* and *Bryum pseudotriquetrum*. And this in particular due to the habitat requirements, much more associated with water in the endemic species [49]. The rapid permanent ice-melting in areas like the South Shetland can result in dryer areas and reduction of plant communities.

Antarctica was the last continent discovered and in the first botanical studies the samples revealed one main taxonomical difficulty: no fertile mosses were found. Among the Antarctic mosses only 22 are found commonly fertile [18] despite some of these species are relatively rare. Reproduction by spores is only possible with water in liquid form, since the antherozoid needs to swim to the correspondent archegonia, fertilize it to form sporophyte and then finely the spores are formed inside a capsule. Since Antarctica is known to have water mostly if form of ice and snow, and being called the biggest desert in the world, it is somewhat difficult and sometimes impossible in some areas to achieve fertilization.

There is a huge difference in precipitation from Dry Valleys at 77,8° S (50 mm) to Livingston Island at 62.6° S (80 mm) [50]. An increase in precipitation was found at Faraday Station, according to data collected from 1956 to 1992 in the Antarctic Peninsula. This increase is connected with the diminishing sea ice and the intensification of evaporation, a higher humidity of the air and more dynamic cyclonic activity, especially in the winter season [51]. All these aspects can affect directly Antarctic plants, contributing also to mosses achieve fertilization.

In tropical and temperate areas ca. 75 ± 90% of the mosses are found fertile. In the maritime Antarctic the value reduces to approximately 25 ± 33% and in continental Antarctica to only 10%. In Margerite Bay fertility of 43% was found (19 species: 17 mosses, 2 liverworts) and in Alexander Island 47% (17 species; 16 mosses, 1 liverwort). 51 sterile species were found among 111 known from Antarctica (46% with sporophyte) [18, 35].

It is interesting that mostly saxicolous mosses are found fertile (**Table 1**), and this is probably because the rock surface is hottest than the environment, melting the snow deposited and resulting in liquid water available more frequently than on other surfaces. As the species usually grow on cracks, the water is piped over them [52].

There is also some preference for the availability of nutrients, especially the presence or proximity to nesting points or with the presence of birds or mammals. These species are called ornithocoprophylous or nitrophilous and often growing on slopes bathed in the excrement of the animals that occur above. Species such as *Synchitria magellanica* and *Henediella heimii* (**Figure 10**), among others, have this preference. With the reduction of penguin populations, already mentioned above, the unavailability of nutrients will affect these species.

Another group is represented by species that do not support high levels of nitrogen and therefore occur away from places where birds or mammals occur. They are called ornithocoprophobic or nitrophobic. Examples are *Pohlia cruda* and *Bartramia patens* (**Figure 9**). These classifications can be used for lichens as well and both mosses and lichens can grow associated in these places.

Mosses can be useful for Antarctic biodiversity, serving as food, as material for making nests or as a resting place for fauna. As food, they are used for this purpose mainly by arthropods, who are permanent residents of the South Pole, as there is no way out of there in winter. In this context, several other groups of microscopic

SPECIES	ON ROCK	ON FINE SEDIMENTS	FERTILITY
Andreaea regularis	X		Frequent
Andreaea gainii		X	Frequent
Schistidium cupulare		X	Rare
Schistidium amblyophyllum		X	Frequent
Schistidium deceptionensis		X	Rare
Schistidium leptoneuron		X	Rare
Schistidium antarctici	X	X	Frequent
Schistidium hialinae		X	Frequent
Schistidium urnulaceum		X	Frequent
Schistidium steerei	Not truly saxicolous	X	Frequent
Schistidium andinum	X	X	Frequent
Schistidium praemorsum		X	Rare
Schistidium rivulare	X	X near water	Frequent
Schistidium lewis-smithii	X	X	Rare
Hymenoloma grimmiaceum		X	Frequent
Hymenoloma crispulum		X	Frequent
Hymenoloma antarcticum		X	Frequent

Table 1.
List of saxicolous/soil growing mosses frequently found fertile in Antarctica.

beings are also inserted, with nematodes or even the smaller rotifers. Another important aspect is the associated microalgae communities.

There is also important associations with large animals, such as birds, which use plants to make their nests. The most used material can be moss (**Figure 12**), there may be mixtures with lichens in different proportions or even with flowering plants, but in some cases lichens (**Figure 14**) and phanerogams may predominate. There are, of course, birds that use other materials, such as rocks, in the case of giant petrels and penguins (**Figure 13**), mud with algae as is the case of *Phalacrocorax atriceps*, etc. [53].

Figure 12.
Skua nest build using the moss Polytrichastrum alpinum *(above) and another using* Sanionia uncinata *(below).*

Figure 13.
*Giant petrel nest (*Macronectes giganteus*) build using rock fragments.*

Figure 14.
Larus dominicanus *(kelp gull) nest build with mosses and the lichen Usnea.*

4. The lichenized fungi: lichens

Lichens are the most representative land group in Antarctica, despite they are not truly plants. They are formed by the symbiosis between a fungus plus an alga (most), a fungus plus a bacterium (case of *Leptogium puberulum,* as for example) or a fungus plus an alga and a bacterium (case of *Placopsis contortuplicata*). There are even lichenized mushrooms such as in *Lichenomphalia* spp. In the relationship, the photobiont provides the carbon source to the fungus, which can be polybasic alcohol (if it is green algae) or glucose (cyanobacteria). The fungus protects the algae from radiation and desiccation. The fungus still manages to reproduce in most cases through sexually formed spores or conidia (asexual), to fragments of the thallus or soredia. The algae reproduction is inhibited or suppressed [54].

To grow like a lichen, the spore needs to find the compatible algae that is rare in nature and lichenize. About 17,500 species of lichenized fungi and about 200 species of associated algae (100 green and 100 cyanobacteria) have been described. In this way, all of these fungi use algae in common and even different algae are used by the same species, in most cases even to adapt better to certain environments [55].

There are approximately between 386 to 427 species of lichens cited for Antarctica [55, 56] numbers that implies the most biodiverse group among terrestrials. In Antarctica in addition to the climate, limiting factors for lichens are the availability of substrate, which in most cases are rocks (in saxicolous species) or mosses (when species are muscicolous) and the presence of a source of nutrients, which can be originating from resting places or breeding animals, as already mentioned in the topic about mosses, above. These species are also starting competition with introduced ones which are being more and more frequent due climatic change.

5. Plant species associated with lichens

In natural environments on the planet a succession is expected to occur. But these environments generally have trees. How is the succession of species in a mainly cryptogamic community like in Antarctica?

Perhaps one of the most ignored formation in Antarctica is that of the lichen/moss association. Mosses colonize an environment first and, to be replaced, must be annihilated. Who does that? If not an animal, mostly a set of lichens. If we look at the work already done with phytosociology in Antarctica, we see that lichens have figured as one of the most important when considering the ecological significance index [17].

Figure 15 illustrates how different species are associated with a lichen which in this case is fruticose: *Sphaerophorus globosus*, which forms groups up to 10 cm in height and is generally parasitic on mosses (muscicolous). In this 20 x 30 square in the figure, there are associated eight other species, of which 3 are mosses and 5 are other lichens, demonstrating how the lichen community settles on mosses and needs them to develop, even if it results in its death. In succession, it is to be expected, therefore, that lichens from the vegetation damage or kill a previously installed moss and then gradually disappear, also due to the lack of a host.

In this community the mosses are at a disadvantage, as they are being attacked by various parasites of the lichen group. These parasites do not even care about the moss species, but it looks like the *Chorisodontium acyphyllum* moss is surviving well and unscathed. This is also noticed in other parts and perhaps indicates that this moss ends up taking the place of the other parasitized and previously killed. This may show a stage of plant succession in Antarctica, which is still poorly studied.

Lichens can also occur on rock fragments and in **Figure 16** there is a schematic drawing of the cover of round rocks, very common in uplifted areas. There are

Figure 15.
Biodiversity surrounding the fruticose lichen Spaerophorus globosus.

Figure 16.
Biodiversity in a 20 x 20 cm square of rounded rocks in Henequin point, King George Island, with 18 species.

17 species of lichens and one of moss occurring on the fragments. At some distance and to the unsuspecting it seems that the rocks have no vegetation, but it has adapted very well to this surface. Vegetation-free areas are in most cases rocks turned over by an animal or researcher who passed the site. So even small areas can group a considerable richness and the Antarctica is very sensitive to any disturbance, imagine the effects of climatic change.

6. Macroscopic terrestrial algae

Prasiola is the macroscopic alga that occurs in terrestrial environments in the Antarctic region with greater frequency. Only two species were being cited for the area: *P. crispa* (nitrophilous) and *P. calophylla* (nitrophobous) (Putzke & Pereira, 2013). Studying the molecular phylogeny of these algae in Antarctica, the presence of *P. crispa* was confirmed, that *P. calophylla* is different from the same species mentioned for Europe, changing its name to *P. glacialis* and that *Prasiola antarctica* is an independent species, morphologically identical *P. crispa* [57].

These species are among the largest primary producers in Antarctica and studies have shown that *P. crispa* is very resistant to desiccation and hypersaline conditions [58–61].

In general, nitrophilic species occur near or inside bird colonies and nitrophobic in areas in contact with them. Often, some shallow pools of water have groups of *Prasiola* that prevent the growth of the surrounding mosses, demonstrating that they are somewhat allelopathic.

In several places it can be seen that the alga is lichenized, forming a different, more blackened and dotted thallus. It is the association with the fungus *Mastodia tesselata*, whose relationship is still controversial, as some authors believe it is parasitism and not a symbiosis. The lichen appears close to Verrucaria, a lichen with marine affinities [62].

In some cases, during the collections it can be seen that part of the algae stem is green and part is associated with the fungus and is already blackened, showing that the association may not be complete. Further studies are needed to elucidate what is the relationship between these two very different organisms.

P. crispa produces secondary metabolites with high toxicity and insecticidal power, and some studies on the subject have already been published [63, 64].

7. Dispersion of land plants

Birds can contribute to the long-distance dispersion of spores and seeds. In the first case, they can carry diaspores of fungi, mosses and pteridophytes transcontinentally (the latter group does not yet occur in Antarctica). Seeds in general can be carried via the digestive tract even. Some first evidences of dispersion of microscopic bryophyte spores have been published, where the case studied presents a transequatorial dispersion, with species carrying diaspores from one pole to the other or at least from the southern part of South America to the North Pole. Algae cells, fragments of moss leaves, elatheria and fungal spores have been found [65]. With the temperature registered due to climate changes, it is expected that new mosses may occur from introductions with the participation of birds in Antarctica.

Many birds carry these structures passively, as they can land on the fields, brush against them, use plants from these groups as material for their nests and even ingest material when transporting the food that is taken to the nests, or when feeding on carcasses arranged on plant communities (**Figures 12–14**). As any fragments of mosses may be sufficient to germinate and form new plants, many species in Antarctica today may have arrived there using transport in the bodies of birds and many more can be introduced in the future.

Feces can also introduce botanical material, as long as it is possible for structures to survive mechanical crushing and chemical bombardment of the digestive tract. The fact that they are eliminated with feces guarantees at least an initial supply of nutrients for their development and, since they are very small plants, the supply deposited once, may even be available for some years.

Despite this, the wind is the most important disperser of Antarctic mosses and lichens, since their main reproductive structures are spores or thallus fragments.

8. The Antarctic ice-free areas and their potential for the evaluation of climatic changes

Antarctica, despite having the largest number of superlatives, as it is considered the coldest (−89.2° C), the driest (average annual precipitation not exceeding 100 mm), the highest (average height 2300 m), the windiest (wind speed can reach 327 Km/h), is the most unknown and the most preserved continent. However, it is isolated from the rest of the world due to its geography and ocean currents. From this isolation the populations are very different, facilitating the study of biological models, whose data can help in the explanation of global biological problems. Global climate change has been a feature in polar regions and continues to be. When discussing climate change on Earth, references are always made to glaciations and ice records [66].

The environmental superlatives of Antarctica, which determine extreme abiotic conditions for the biota, led to the evolution of fragile and unique communities, which are mainly characterized have high specialization and adaptation to environmental conditions, in addition to being very sensitive to environmental impacts of anthropic origin or caused by natural phenomena.

The climatic phenomena that occur in Antarctica are the basis for describing the climate of the Southern Hemisphere, and what happens in many countries is in part also a reflection of the phenomena that occur in the South Pole. In Antarctica the so-called "fronts" are frequent, numerous and of constant formation, these are mostly ephemeral, but many reach the southern areas of South America. In addition, the Antarctic ice is considered as a climatic archive. Air bubbles found in glacial ice can identify the composition of air from past eras. Snow samples can currently demonstrate the types of gases and particles that existed in atmospheric air many years ago. This means that through isotopes, it is possible to evaluate the activity of the sun in several eras, in addition to the biological activity, obtained by the analysis of molecules of organic origin [66].

Pollution was believed to be almost exclusively a product of the Industrial Age, but ice samples demonstrated lead pollution, dating from the Roman Empire period. The snow when deposited carries with it the characteristics of the chemical composition of the atmosphere at the moment it was formed, deposited on the continent's surface air bubbles, salts, dust, volcanic ash, pollutants, among others. As snow does not melt on glaciers, the layers are deposited and compacted, keeping the record of climatic phenomena that occur over time preserved.

One of the global changes that can affect the Antarctic ecosystems is the so-called "hole in the Ozone layer", which is located at the south pole, because it brings together the coldest regions of the planet and for having a very localized circulation of air masses. This despite being on Antarctica also reaches the southern tip of South America. This phenomenon contributes to the increase in ultraviolet radiation (UV-B), which, because it is mutagenic, contributes to the genetic alteration of species. Since this radiation is very intense in Antarctica, great mutation rates are expected, but it was observed that this does not occur, since these species have mechanisms that prevent DNA damage by the formation of secondary metabolites, at least in plants, whose photoprotector effects were experimentally proven [25, 26].

The retreat of the glaciers and the reduction of snow fields expand and the consequent exposure of new habitats for colonization, and the increase in the populations of plants, has been documented. Small changes in the physiology of the Antarctic organisms can affect their life histories, with indirect effects on the dynamics of the ecosystem and trophic chains. These subtle effects can be more easily detected due to the simplicity of polar ecosystems [25].

The use of plant communities in Antarctic ice-free areas to assess climatic changes consequences can be justified by facts such as:

1. These have a small number of species when compared to periglacial and subtropical regions, since among the species mentioned and described so far there are: two species of Magnoliophyta, *Deschampsia antarctica* Desv. (Poaceae) and *Colobanthus quitensis* (Kunth.) Bart. (Caryophyllaceae), approximately 360 species of lichens [55]. Bryophytes comprise approximately 113 species of moss and 22 species of liverworts [18].

2. As biodiversity is small, populations are very numerous, facilitating their delimitation and the identification of interspecific relationships.

3. The presence of soil is an important factor, since there are species of mosses such as, for example, *Sanionia uncinata* (Hedw.) Loeske and *Chorisodontium aciplhyllum* (Hook. F et Wils.) Broth. that grow in areas where rock fragments occur, as the soil is formed these populations are replaced by other moss species such as, for example, *Polytrichum juniperinum* Hedw. and *Polytrichastrum alpinum* (Hedw.) G.L.Smith, often associated with *Deschampsia antarctica* and *Colobanthus quitensis* [45].

4. Most of the species that grow in these areas, evolved under extreme environmental conditions and under intense stress, making them very well adapted to such environmental conditions.

5. Antarctica is still a continent with an insignificant anthropic impact, so the changes that occur in communities are the result of environmental changes arising from natural phenomena. This fact is important, since environmental variables that are selected can be evaluated based on natural phenomena.

6. The importance of studying plant species that grow in ice-free areas in Antarctica are strongly related to the environment, so it constitutes a potential source for assessing global changes. It is expected that climate change will have a major impact on Antarctic land biota. Studies suggest that the increase in temperature and greater availability of water can extend periods favorable to growth, increase the rates of development and reduce the duration of the life cycle, which can alter the distribution of species [5].

Acknowledgements

We acknowledge the Brazilian Antarctic Program and the CAPES and CNPq for financial support and logistic facilities to field work in Antarctica.

Conflict of interest

The authors declare no conflict of interest.

Author details

Jair Putzke* and Antonio Batista Pereira
Federal University of Pampa, São Gabriel, Brazil

*Address all correspondence to: jrputzkebr@yahoo.com

References

[1] Newsham KK, et al. Relationship between soil fungal diversity and temperature in the maritime Antarctic. Nature Climate Change. 2016;6:182-186.

[2] Rignot E, et al. Ice-shelf melting around Antarctica. Science. 2013;341:266-270.

[3] IPCC. Intergovernmental Panel on Climate Change. 2007. Climate Change 2007: Synthesis Report. Contribution of Working Groups I, II and III to the Fourth Assessment. IPCC, Geneva, Switzerland.

[4] Monaghan AJ, Bromwich DH, Chapman W, Comiso JC. Recent variability and trends of Antarctic near-surface temperature. Journal of Geophysical Research. 2008.113(D04105):114.

[5] Turner J, Colwell SR, Marshall GJ, Lachlan-Cope TA, Carleton AM, Jones PD, Lagun V, Reid PA, Iagovkina S. Antarctic climate change during the last 50 years. Journal of Climatology. 2005;25:279-294.

[6] Rosa KK, Vieira R, Fernandez G, Mendes CWJr, Velho LF, Simões JC. Recent changes in the Wanda Glacier, King George Island, Antarctica. Pesquisas em Geociências. 2015;42(2):187-196.

[7] Joughin I, Smith BE, Medley B. Marine ice sheet collapse potentially under way for the Thwaites Glacier Basin, West Antarctica. Science. 2014;344:735-738.

[8] Cook AJ, et al. Ocean forcing of glacier retreat in the western Antarctic Peninsula, Science. 2016; 283-285. doi:10.1126/science.aae0017

[9] Greenbaum JS, et al. Ocean access to a cavity beneath Totten Glacier in East Antarctica. Nature Geoscience. 2015;8:294-298.

[10] Amesbury MJ, et al. Widespread Biological Response to Rapid Warming on the Antarctic Peninsula. Current Biology. 2017;27:1616-1622.

[11] Gray A, et al. Remote sensing reveals Antarctic green snow algae as important terrestrial carbon sink. Nature Communications. 2020;11:1-9.

[12] Xavier-Filho L, Guzman GG, Rudolph ED. Observations on nutrient flow in a maritime Antarctic Tundra Community. Boletim da Sociedade Broteriana. 1989;62:97-106.

[13] Ainley D, Russell J, Jenouvrier S, Woehler E, Lyver PO, et al. Antarctic penguin response to habitat change as Earth's troposphere reaches 2°C above preindustrial levels. Ecol. Monogr. 2010;80:49-66.

[14] Wilson P, Ainley D, Nur N, Jacobs S, Barton G, et al. Adélie penguin population change in the pacific sector of Antarctica: relation to sea-ice extent and the Antarctic Circumpolar Current Marine Ecology Prog. Ser. 2001;213: 301-309.

[15] Stonehouse B. The biology of penguins. London: University Park Press. 1975.

[16] Cimino MA. et al. Projected asymmetric response of Adélie penguins to Antarctic climate change. Scientific Reports 2016. 6: 28785.

[17] Schmitz D, Putzke J, Albuquerque MP, Schünneman AL, Vieira FCB, Victoria FC, Pereira AB. Description of plants communities from Half Moon Island, Antarctica. Polar Research. 2018;37:1-12.

[18] Ochyra R, Lewis-Smith RI, Bednarek-Ochyra H. The illustrated moss flora of Antarctica. Cambridge University Press. 2008. 685 p.

[19] Lewis-Smith RI. Destruction of Antarctic terrestrial ecosystems by rapidly increasing fur seal population. Biological Conservation. 1988;45:55-72.

[20] Lewis-Smith RI. Signy Island as a paradigm of biological and enviromnental change in Antarctic terrestrial ecosystems. In: K. R. Kerry & G. Hempel (eds). Antarctic Ecosystems: ecological change and conservation. Berlin. Springer-Verlag: 1990. pp. 32-50.

[21] Lewis-Smith RI. Impact of an increasing fur seal population on Antarctic plant communities: resilience and recovery. In: B. Battaglia, J. Valencia & D. W. H. Walton (eds.). Antarctic communities: species, structure and survival. Cambridge. Cambridge University Press: 1997. pp. 432-436.

[22] Pereira AB, Putzke MTL, Putzke J. Biological Communities of Keller Peninsula, King George Island - Antarctica. Caderno de Pesquisa. Série Biologia. 2008;20:63-74.

[23] Zuñiga GE, Alberdi M, Corcuerm LJ. Non-structural carbohydrates in *Deschampsia antarctica* Desv. from South Shetland Islands, Maritime Antarctic. Environmental and Experimental Botany. 1996;36(4):393-399.

[24] Casanova-Katny MA, Cavieres LA. Antarctic moss carpets facilitate growth of *Deschampsia antarctica* but not its survival. Polar Biology. 2012;35:1869-1878.

[25] Kappel Pereira B, Pereira AB, Groff A, Schroder-Pfeifer NT, Silva J. Avaliação do efeito fotoprotetivo de Polytrichum juniperinum Hedw., Colobanthus quitensis (Kunth.) Bartl. e *Deschampsia antarctica* Desv., através de Ensaio Cometa em *Helix aspersa* (Mülller, 1774). In: Actas do V Simposio Argentino e I Latinoamericano sobre Investigaciones Antarticas. 2004;1:1-4.

[26] Kappel Pereira B, Rosa RM, Silva J, Guecheva TN, Oliveira HM, Lanistcki M, Benvegnú VC, Furtado GV, Ferraz A, Richter MF, Schoder N, Pereira AB, Henriques JAP. Protective effects of three extracts from Antarctic plants against ultraviolet radiation in several biological models. Journal of Photochemistry and Photobiology. B. Biology. 2009;96:117-129.

[27] Day TA, Ruhland CT, Xiong FS. Warming increases aboveground plant biomass and C stocks in vascular-plant dominated Antarctic tundra. Global Change Biology. 2008;14:1827-1843.

[28] Day TA, Ruhland CT, Strauss SL, Park JH, Krieg ML, Krna MA, Bryant DM. Response of plants and the dominant microarthropod, *Cryptopygus antarcticus*, to warming and contrasting precipitation regimes in Antarctic tundra. Global Change Biology. 2009;15:1640-1651.

[29] Fowbert JA, Smith RIL. Rapid population increases in native vascular plants in the Argentine Islands, Antarctic Peninsula. Arctic Alpine Research. 1994;26:290-296.

[30] Cannone N, Guglielmin M, Convey P, Worland MR, Favero-Longo SE. Vascular plant changes in extreme environments: effects of multiple drivers. Climatic Change. 2016;134:651-665.

[31] Torres-Mellado GA, et al. Antarctic hairgrass expansion in the South Shetland archipelago and Antarctic Peninsula revisited. Polar Biology. 2011;34(11):1679-1688.

[32] Putzke J, Pereira AB. Phaeosphaeria deschampsii (Ascomycota): A new parasite species of *Deschampsia antarctica* (Poaceae) described to Antarctica. Anais da Academia Brasileira de Ciências (Online). 2016;88 (3):01-03.

[33] Velazquez D, Lopez-Bueno A, Aguirre De Carcer D, De Los Rios A, Alcami A, Quesada A. Ecosystem function decays by fungal outbreaks in Antarctic microbial mats. Scientific Reports. 2016;6: doi:10.1038/srep22954.

[34] Quintana RD, Benitez VCO. Nest material of skuas (Catharacta spp.) and kelp gulls (*Larus dominicanus*) at Cierva Point, Antarctic Peninsula. Notornis. 2001;48:235-241.

[35] Convey P, Smith RIL. Responses of terrestrial Antarctic ecosystems to climate change. Plant Ecology. 2006;182:1-10.

[36] Day TA, Ruhland CT, Grobe CW, Xiong F. Growth and reproduction of Antarctic vascular plants in response to warming and UV radiation reductions in the field. Oecologia. 1999;119:24-35.

[37] Putzke J, Ferrari FR, Shaefer CEGR. Discovery of a large population of Hygrolembidium isophyllum (Lepidoziaceae, Marchantiophyta) in the South Shetland Islands, Antarctica. Polar Research. 2020;39:1-4.

[38] Pereira AB, Francelino MR, Roesch LFW. Vegetal communities from ice-free areas of Copacabana, King George Island, Antarctica. Annual Activity Report. 2009; 25-26.

[39] Putzke J, Pereira AB, Putzke MTL. New Record of Myxomycetes to the Antarctica. In: Actas del V Simposio Argentino y I Latinoamericano de Investigaciones Antarticas. 2004;1:1-4.

[40] Newsham KK, Structural changes to a mycothallus along a latitudinal transect through the maritime and sub-Antarctic. Mycorrhiza. 2011; 21:231-236.

[41] Lewis-Smith RI. Plant colonization response to climate changes in the Antarctica. Folia Fac. Sci. nat. univ. Masarykianae Brunensis, Geográfica. 2001;25:19-33.

[42] Newsham KK. Structural changes to a mycothallus along a latitudinal transect through the maritime and sub-Antarctic. Mycorrhiza. 2011;21 (3):231-236.

[43] Roads E, Longton RE, Convey P. "Millennial timescale regeneration in a moss from Antarctica". Current Biology. 2014;24(6):222-223.

[44] Cannone N, Corinti T, Malfasi F, Gerola P, Vianelli A, Vanetti I, Zaccara S, Convey P, Guglielmin M. Moss survival through in situ cryptobiosis after six centuries of glacier burial. Scientific Reports. 2017;7:1-7.

[45] Pereira AB, Putzke J. Floristic Composition of Stinker Point, Elephant Island, Antarctica. Korean Journal of Polar Research. 1994;5(2):37-47.

[46] Putzke J. Bryophytes and their associates in South Shetland Islands – Antarctica. In: Alan Afroz (ed.). Recent Advances in Botanical Science. 2020. p. 8-19.

[47] Royles J, et al. Plants and Soil Microbes Respond to Recent Warming on the Antarctic Peninsula. Current Biology. 2013;23(17):1702-1706.

[48] Clarke LJ, Robinson SA, Hua Q, Ayre DJ, Fink D. Radiocarbon bomb spike reveals biological effects of Antarctic climate change. Global Change Biology. 2012;18:301-310.

[49] Robinson S, Clarke L. Plants and Soil Microbes Respond to Recent Warming on the Antarctic Peninsula. Australian Antarctic Magazine. 2008; 14.

[50] Sancho LG, Pintado A, Green TGA. Antarctic Studies Show Lichens to be Excellent Biomonitors of Climate Change. Diversity. 2019;11(42):1-14.

[51] Turner J, Colwell SR, Harangozo S. Variability of precipitation over the coastal western Antarctic Peninsula from synoptic observations. Journal of Geophysical Research. 1997;102 (D12):13999-14007.

[52] Lewis-Smith RI, Convey P. Enhanced sexual reproduction in bryophytes at high latitudes in the maritime Antarctic. Journal of Bryology. 2013;24:107-117.

[53] Albuquerque MP, Victoria FC, Schünemann AL, Putzke J, Gunski RJ, Seibert S, Petry MV, Pereira AB. Plant Composition of Skuas Nests at Hennequin Point, King George Island, Antarctica. American Journal of Plant Sciences, 2012;3:688-692.

[54] Nash TH. Lichen biology. 2nd ed. Cambridge UK: Cambridge University Press. 2008.

[55] Øvstedal DO, Lewis-Smith RI. Lichens of Antarctica and South Georgia – A guide to their identification and ecology. Studies in Polar Research. Cambridge University Press. 2001. 411p.

[56] Engelen A, Convey P, Ott S. Life history strategy of Lepraria borealis at an Antarctic inland site, Coal Nunatak. Lichenologist. 2010;42(3):339-346.

[57] Moniz MBJ. et al. Molecular Phylogeny of Antarctic Prasiola (Prasiolales, Trebouxiophyceae) Reveals Extensive Cryptic Diversity. Journal of Phycology. 2012;48(4):940-955.

[58] Jacob A. Physiologie und Ultrastruktur der antarktischen GrOnalge Prasiola crispa Ss*p. antarctica* unter osmotischem StreD und Austrocknung. Berichte zur Polarforschung. 1992;102:1-143.

[59] Jacob A, Kirst GO, Wiencke C, Lehmann H. Physiological responses of the Antarctic green alga Prasiola crispa Ss*p. antarctica* to salinity stress. Journal of Plant Physiology. 1991;139:57-62.

[60] Jacob A, Lehmann H, Kirst GO, Wiencke C. Changes the ultrastructure of Prasiola crispa under salinity stress. Botanica Ada. 1992a;105:41-46.

[61] Jacob A, Wiencke C, Lehmann H, Kirst GO. Physiology and ultrastructure of desiccation in the green alga Prasiola crispa from Antarctica. Botanica Marina. 1992b;35:297-303.

[62] Pérez-Ortega S, Ríos Ade L, Crespo A, Sancho LG. Symbiotic lifestyle and phylogenetic relationships of the bionts of Mastodia tessellata (Ascomycota, incertae sedis). American Journal of Botany. 2010;97(5):738-752.

[63] Dos Santos SM, Ramos CJB, Leite JPG, Teixeira VL, De Palmer Paixão ICN, Belo CA, Pereira AB, Pinto AMV. Antiviral activity of 7-keto-stigmasterol obtained from green Antarctic algae Prasiola crispa against equine herpesvirus 1. Journal of Applied Phycology. 2016;28:1-8.

[64] Zemolin APP, Cruz LC, Paula MT, Pereira BK, Albuquerque MP, Victoria FC, Pereira AB, Posser T, Franco J. Toxicity Induced by Prasiola crispa to Fruit Fly *Drosophila melanogaster* and Cockroach *Nauphoeta cinerea*: Evidence for Bioinsecticide Action. Journal of Toxicology and Environmental Health. 2014;77:115-124.

[65] Lewis-Smith RI. et al. First evidence of bryophyte diaspores in the plumage of transequatorial migrant birds. PeerJ. 2014;2: e424 pp. 1-13.

[66] Brito TAS. Antártica bem comum da humanidade. Ministério do Meio Ambiente. 2009. 69 p.

Chapter 5

Whales as Indicators of Historical and Current Changes in the Marine Ecosystem of the Indo-Pacific Sector of the Antarctic

Yoshihiro Fujise and Luis A. Pastene

Abstract

We review the scientific information on whales that could be indicative of historical and current changes in the ecosystem in the Indo-Pacific sector of the Antarctic. The increased krill availability in the middle of the past century as a result of the heavy harvesting of the larger baleen whale species could have been translated into better nutritional conditions for the Antarctic minke whale, resulting in a decreasing trend in the age at sexual maturity and an increasing trend in recruitment rate and hence total population size between approximately 1940 and 1970. This nutritional condition has deteriorated more recently, as revealed by a decrease in energy storage and stomach content weight since the 1980's; these changes coincide with appreciable increases in the abundances of humpback and fin whales, which were heavily harvested in the first half of the past century. The historical demographic changes observed in the Antarctic minke whale are consistent with the pattern to be expected under the krill surplus hypothesis, with minke whales now again competing with other (recovering) baleen whale species for krill. However, these minke whales could also be using alternative feeding areas (e.g. polynias within the pack-ice) in response to the increase in abundance and geographical expansion of these other large whale species. This could provide an alternative explanation for indications from sighting surveys and population models of a decrease and then re-stabilisation of minke whale abundance in open water areas since the 1970s.

Keywords: East Antarctic, Antarctic minke whale, blue whale, fin whale, humpback whale, whaling, krill surplus hypothesis, abundance, biological parameters, nutritional condition

1. Introduction

The Antarctic ecosystem is very dynamic with changes in species composition and habitat occurring through time. Within this ecosystem, Antarctic krill

(*Euphausia superba*) is a key prey species, supporting different species of baleen whales, pinnipeds, birds and fish. Changes in the ecosystem can result from human interventions or from natural causes. One example of human intervention is the large-scale harvesting of whales in the first half of the 20th Century, which has been discussed by several authors [1, 2]. This harvesting started in the Antarctic Ocean in 1904. Several species of krill-eating large whales, such as the Antarctic blue (*Balaenoptera musculus intermedia*) and humpback (*Megaptera novaeangliae*) whales were heavily reduced in number by commercial whaling during the first half of the past century. Other species such as the fin whales (*B. physalus*) were reduced during the second half. Over more recent decades, the populations of some large whales have started to recover [3]. Changes in the biomass of whale species also seem to have had strong effects on the demography of other krill-eating predators in the Antarctic ecosystem [1, 2].

An example of the effects of natural causes is the increases in the chinstrap penguins (*Pygoscelis antarctica*) populations of the Scotia and Weddell Seas over the last four decades (1950's-1990's), which has been attributed to a gradual decrease in the frequency of cold years with extensive winter sea ice cover resulting from environmental warming [4]. However, more recent analyses in the Antarctic Peninsula and Scotia Sea conclude that the chinstrap penguin instead may be among the most vulnerable species affected by a warming climate [3].

In studying the changes in the Antarctic ecosystem, there needs to be differentiation between West and East Antarctic, as well between land-based and sea-based krill predators. The West Antarctic Peninsula represents one of the regions of the planet where the fastest warming has been observed in the last 50 years [5]. For this reason the studies documenting ecosystem changes in the West Antarctic have considered environmental variables in addition to demographic information on land-based krill predators (mainly penguin species) [3, 4], on which environmental factors could have a larger impact. Warming has not been reported for the East Antarctic, so that environmental factors would not be expected to play the predominant role in the ecosystem changes in this part of the Antarctic.

Here historical and current ecosystem changes in the Indo-Pacific sector of the Antarctic (involving mostly East Antarctic) are documented through the examination of biological and demographic parameters of sea-based predators (whales). These changes in parameter values are interpreted in the context of some established hypotheses.

2. Characterization of the research area

The present study is focused on the Indo-Pacific sector of the Antarctic, in the longitudinal range 35°E-145°W (**Figure 1**), south of 60°S, which is the approximate position of the Antarctic Convergence (AC). This longitudinal range includes International Whaling Commission (IWC) management Areas III (east part), IV, V and VI (west part) (**Figure 1**).

Prydz Bay is located at the west boundary of the research area while the Ross Sea is located at its east boundary. The research area is strongly influenced by the southern boundary of the Antarctic Circumpolar Current (SBACC), which interacts with the coastal East Wind Drift (EWD) in a series of fronts and eddies (**Figure 1**). A series of gyres link the EWD and the SBACC, e.g. the Prydz Bay and the Ross Sea gyres. Krill concentrations appear to track gyral systems off the East Antarctic, for example in the sectors between 30°-90°E or 80°-115°E [6].

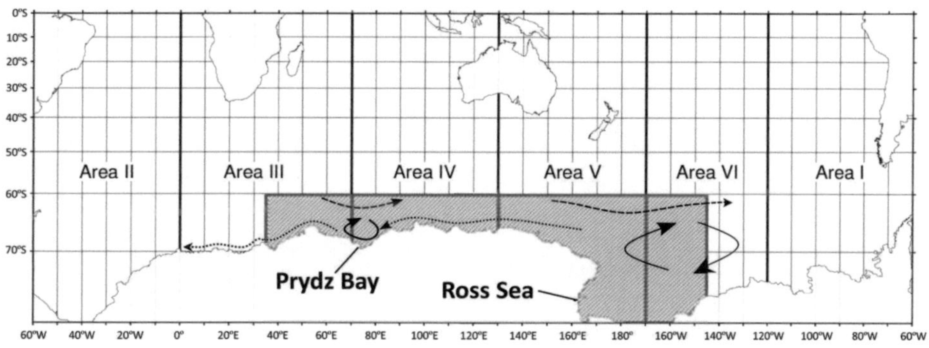

Figure 1.
Schematic representation of the research area (dashed in blue). The figure shows the Southern Hemisphere Areas that the International Whaling Commission (IWC) uses for the management and conservation of baleen whales (except the Bryde's whale, B. edeni). The research area is influenced by the southern boundary of the Antarctic Circumpolar Current (SBACC) (dashed arrow), which interacts with the coastal East Wind Drift (EWD) (dotted arrow) in a series of fronts and eddies.

3. Krill-eating baleen whale species in the Indo-Pacific sector of the Antarctic

Baleen whale species, except the Bryde's whale, migrate seasonally between low latitude breeding areas in winter to high latitude feeding areas in the Antarctic in summer. The main prey species of baleen whales such as Antarctic blue, fin, humpback and Antarctic minke (*B. bonaerensis*) whales (**Figure 2**) is the Antarctic krill. Therefore the summer migrations of these whales to the Antarctic are related to areas of krill concentrations, which in turn are associated with gyral systems.

3.1 Antarctic blue whale

This is the largest baleen whale species. The record for a whale killed in the Southern Hemisphere in the first half of the past century was a body length of more than 30 m and weigh of nearly 180 tons. During the austral summer Antarctic blue whales are distributed between the AC and the ice edge. There is limited information on the population structure of this species.

3.2 Fin whale

This is the second largest baleen whale species, with a maximum length of more than 27 m and weight of nearly 120 tons. During the austral summer, fin whales are found extensively south of 50°S, but most commonly north of 60°S. There is limited information on the population structure of this species.

3.3 Humpback whale

This species presents a maximum body length of 17 m and weight of 40 tons. During the austral summer humpback whales are distributed from south of the AC to the ice edge, but just to the north of the main distribution area for Antarctic minke whale. The IWC has identified seven populations of humpback whales in the Southern Hemisphere, which are denominated with alphabetic letters from 'A' to 'G' [7]. The populations occurring in the Indo-Pacific sector of the Antarctic are Populations 'C' (mainly in Area III), 'D' (mainly in Area IV), 'E' (mainly in Area V), and 'F'

Figure 2.
Krill-eating baleen whale species in the Indo-Pacific sector of the Antarctic. Top left: humpback whale; top right: Antarctic minke whale; bottom left: fin whale; bottom right: Antarctic blue whale.

(mainly in Area VI) (**Figure 1**). There are some spatial overlaps between adjacent populations in the Antarctic [8]. The breeding areas for Populations 'D' and 'E' are located in West and East Australia, respectively.

3.4 Antarctic minke whale

This is one of the smallest baleen whale species with a maximum body length of more than 10 m and weight of nearly 10 tons. During the austral summer, Antarctic minke whales are distributed mainly around the pack-ice. There are at least two populations of this species in the Indo-Pacific sector of the Antarctic, the Eastern Indian Ocean population in the western part of the research area (mainly the eastern part of Area III and Area IV) (**Figure 1**), and the Western South Pacific population in the eastern part of the research area (mainly the eastern part of Area V and Area VI) (**Figure 1**) [9, 10]. Both populations interact in a transition area between approximately 100° and 160°E (eastern part of Area IV and western part of Area V) (**Figure 1**) [11].

The biological and demographic studies summarized below are based on the approximate geographic limits of these 'populations' in the case of humpback and Antarctic minke whales, and on geographical areas for those species with limited information on population structure (Antarctic blue and fin whales).

4. Whale and environmental surveys in the Indo-Pacific sector

As explained briefly above, the causes of ecosystem changes in the Antarctic are complex. To determine those causes, long-term monitoring research programs focused on collecting biological data of krill predators, as well data on sea ice cover and environmental variables, are important. The kind of information which is used to monitor changes in the ecosystem through sea-based predators (whales) and their environment is shown in **Table 1**.

Most of the data in **Table 1**, which have been used in the studies summarized below, come from two main sources.

Parameter	How the information is obtained?	Relevance of monitoring
Krill biomass	Echo-sounder and net surveys	Krill is a key species in the Antarctic ecosystem. Changes in its abundance have effects on predators and the whole ecosystem
Whale abundance	Systematic sighting surveys	Fluctuations in the abundance of whales over time is important for their management. Different levels of whale abundance have different impacts on krill
Whale distribution	Systematic sighting surveys	Distributions of whale species can change with time in response to changes in abundance and/or changes in oceanographic conditions/krill abundance
Whale recruitment	Population dynamic models that use age and abundance information for whales	Same as above. Index of young whale abundance
Blubber thickness, fat weight and girth	Direct measurements from sampled whales	Index of body condition. Better nutritional condition (e.g. better availability through higher abundance of krill) will be reflected in thicker blubber, heavier fat and larger girth
Stomach content weight	Direct measurements from sampled whales	Index of body condition. Better nutritional condition will be reflected by heavier stomach contents
Age at sexual maturity (ASM)	Examination of the transition phase in earplugs; examination of ovaries and testis	Better nutritional conditions will be reflected in a shift of the ASM to younger ages, so that whales will be able to reproduce at younger ages
Pregnancy rate	Examination of ovaries and uterus	Better nutritional conditions will be reflected in higher pregnancy rates
Oceanographic conditions	Systematic oceanographic surveys based on CTD and XCTD	Changes in oceanographic conditions will affect the distribution and krill biomass, and in turn the abundance and distribution of whales. Changes in oceanographic conditions might be an effect of climate change

Table 1.
Biological and ecological parameters monitored for whales and their environment to investigate changes in the ecosystem.

4.1 JARPA and JARPAII programs

The Japanese Whale Research Program under Special Permit in the Antarctic (JARPA) was conducted in the austral summers from 1987/88 to 2004/05, and its second phase (JARPAII) from 2005/06 to 2013/14. Both programs conducted systematic surveys in the Indo-Pacific sector (35°E-145°W) of the Antarctic using both lethal (biological sampling of a limited number of Antarctic minke whales) and non-lethal (biopsy sampling and photo-identification of large whales, oceanographic and marine debris surveys, dedicated sighting surveys) approaches. The main objectives of these programs were related to the investigation of stock structure, biological parameters and feeding ecology of Antarctic minke whales, as well the investigation of environmental pollutants in whale tissue and the environment. These surveys were conducted in the open sea because the survey vessels were not ice strengthened. Data and research outputs from JARPA and JARPAII were reviewed by IWC-organized international workshop of specialists, and they are available in conjunction with the reports of those workshops [9, 12].

4.2 IWC's IDCR/SOWER programs

The IWC's International Decade for Cetacean Research (IDCR) undertook a series of Antarctic sighting cruises for assessment of Antarctic minke whales during the austral summers 1978/79–1995/96. From 1995/96 this was renamed the Southern Ocean Whale and Ecosystem Research (SOWER) program, and continued until the 2009/10 season. The primary aim of these programs was to conduct sighting surveys using the line transect method for estimating the abundance of Antarctic minke whales and other cetacean species. The survey programs have also enabled collection of biopsies, photo-identification, oceanographic and acoustic samples for studies on the ecosystem. As for JARPA and JARPAII, these surveys were conducted in the open sea because the survey vessels were not ice strengthened. Even though IDCR/SOWER surveys were conducted at a circumpolar level, it is the information from the surveys conducted in the Indo-Pacific sector, particularly in IWC Areas IV and V (see **Figure 1**), that is summarized here.

Some of the studies summarized below used biological data collected during former commercial whaling operations (by Japan and the former USSR) in the Indo-Pacific sector of the Antarctic.

5. Historical ecosystem changes revealed through whale demography

5.1 Trend in age at sexual maturity

Changes in the age at sexual maturity (ASM) indicate changes in the nutritional conditions for the whales, which in turn could indicate less or more food availability in the environment. Better nutritional conditions will be reflected by a shift of the ASM to younger ages, so that whales will be able to reproduce at younger ages and as a consequence the populations will grow faster. One of the methods for determining ASM in whales is through the examination of the 'transition phase' in the earplugs [13]. The earplugs of several baleen whale stocks exhibit seasonal growth layers which have been shown for some species to indicate the age of the animals. A transition from early, irregular layers to later, more regular layers can be seen in these earplugs (the 'transition phase'), and this has been shown to indicate the age at sexual maturity of the whale [13]. Historical changes in the ASM can be investigated through the analyses of cohorts (groups of whales born in the same year).

Earplugs of Antarctic minke whales were collected during the period of commercial whaling in the early 1970's, and during the JARPA/JARPAII surveys in the Indo-Pacific sector of the Antarctic for more than 25 years. A decline in the average age at transition in Antarctic minke whales in the Eastern Indian Ocean population from roughly 11 years for the cohorts of the 1950's to roughly 7 years for the cohorts of the 1970's was evident (**Figure 3**), and this trend was similar for females and males. The ASM remained stable until the 1980's cohorts [14]. This work was subsequently updated [15] by using a large number of samples, and those authors confirmed that the ASM remained stable until the 1998 cohort at least. The same pattern was observed for the Western South Pacific population.

5.2 Trend in recruitment rate and total population size

The Scientific Committee (SC) of the IWC has been applying statistical catch-at-age (SCAA) analyses to Antarctic minke whales since 2005. SCAA is a common method of fisheries stock assessment where age-structured catch data from a

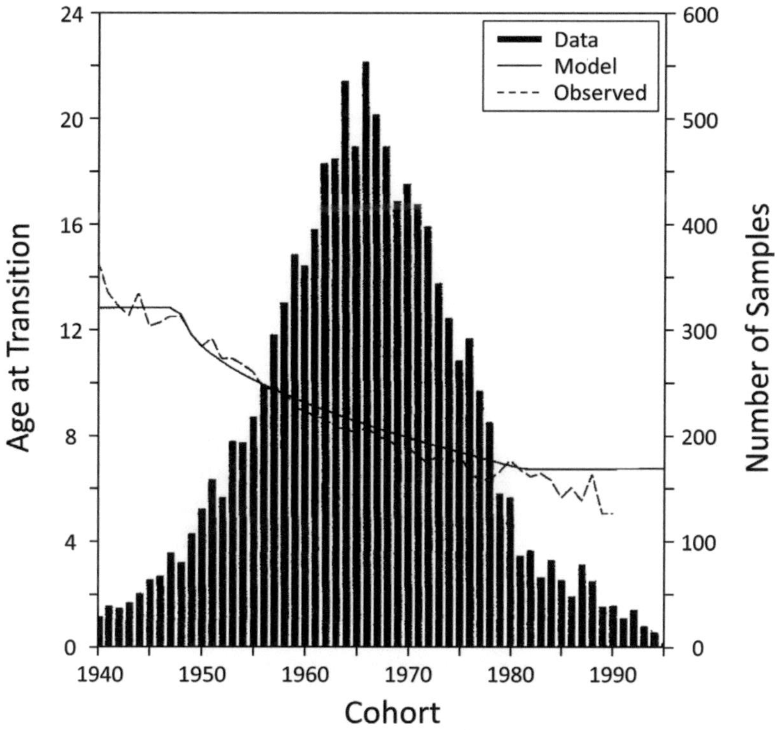

Figure 3.
Changes in the age at sexual maturity of Antarctic minke whale as determined from the transition phase, by cohort (Eastern Indian Ocean population). Histogram of the number of whales aged in each cohort is also showed. Age at sexual maturity changed from around 11 years in the 1950s cohorts to around 7 years in the 1970s cohorts (modified from [14]).

fishery are used to estimate quantities of interest, such as population size and natural mortality rates, employing the maximum likelihood estimation approach [16]. A summary history of the application of SCAA to Antarctic minke whale is provided in [17], and an assessment of this species using SCAA is reported in [18].

The data used when conducting assessment by SCAA on Antarctic minke whales consisted of catches, abundance estimates, length frequency data, and conditional age-at-length data. Different series of abundance estimates were used, i.e. those from the IWC's IDCR/SOWER and JARPA/JARPAII's dedicated sighting surveys in the Indo-Pacific sector of the Antarctic. The biological data mentioned above were available from the period of commercial whaling (Japan and the former USSR), and JARPA/JARPAII surveys in the Indo-Pacific sector of the Antarctic for more than 25 years.

The SCAA assessment on Antarctic minke whale included a 'reference' case and several sensitivity tests. These tests explored sensitivity to the weight assigned to the various data sources and penalties in the model fitting process, to assumptions related to vulnerability, natural mortality and catchability, and to the use or otherwise of the JARPA/JARPAII's abundance index data [18]. As in the estimation of the ASM, historical changes in recruitment and total population size can be investigated through the analyses of cohorts.

Figure 4 shows the temporal trend for the total size of the Eastern Indian Ocean population of Antarctic minke whales. Results presented here refer to the 'reference' case, and were robust to the sensitivity tests conducted. The population was estimated to have increased from 1930 until the early 1970's, with the population having declined subsequently and then staying stable. The increase in abundance

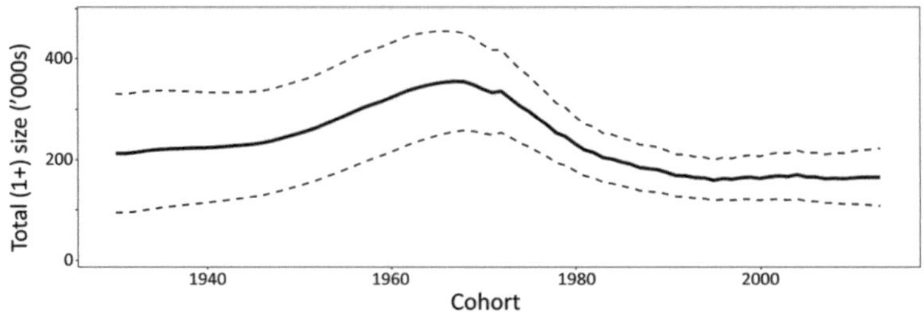

Figure 4.
Time trajectory of total (1+) population size for the Eastern Indian Ocean population of Antarctic minke whale (reference-case), by cohort. The dotted lines indicate 95% asymptotic confidence intervals (modified from [18]).

was due primarily to an increase in recruitment (**Figure 5**), arising in turn from an increase in carrying capacity [18], presumably due to an increase in the abundance of krill. A similar pattern was found for the Western South Pacific population.

It is interesting to note that the increase in total population size from 1930 to the 1970's coincided roughly with the period over which the ASM decreased.

5.3 Interpretation of results

The results of this section can be summarized as following:

- The average ASM of Antarctic minke whales declined from roughly 11 years for the cohorts of the 1950's to roughly 7 years for the cohorts of the 1970's; this trend was similar for females and males, and for the two populations occurring in the Indo-Pacific sector.

- The population of Antarctic minke whale increased from 1930 until the early 1970's, but declined subsequently and then re-stabilised. A similar pattern is evident for the two populations occurring in the Indo-Pacific sector.

- The increase in abundance was due primarily to higher recruitments resulting from an increase in carrying capacity with more food being available for these Antarctic minke whales during the middle decades of the last century.

These historical changes in the demography of Antarctic minke whale are consistent with expectations under the krill surplus hypothesis. The harvesting of large whales in the Antarctic Ocean started in 1904, and several species of large whales such as the Antarctic blue and humpback whales were heavily depleted by

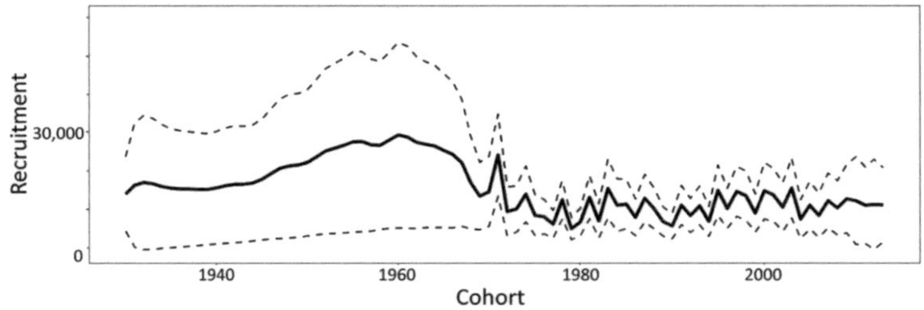

Figure 5.
Time trajectory of recruitment for the Antarctic minke whale of the Eastern Indian Ocean population (from 1930 and from 1975) for the reference case analysis, by cohort (modified from [18]).

the first half of the past century. Other species such as the fin whales were depleted during the second half of that century. Antarctic blue, fin and humpback whales were reduced to 5%, 21% and 2% of their original total sizes of 220,000, 490,000 and 130,000, respectively [19]. However, commercial harvesting of the Antarctic minke whales started only in the early 1970's, when the other baleen whale species were already depleted.

Some researchers have suggested that following the period of heavy harvesting of the large baleen whales in the Antarctic mainly during the middle decades of the past century, some 150 million tons of 'surplus' annual production of Antarctic krill became available for other krill predators, such as Antarctic minke whales, crabeater seals, fur seals, penguins and some albatrosses. These species then took advantage of this food surplus to increase their abundance. This is the so-called krill surplus hypothesis [1, 2].

The increased krill abundance around the middle of the past century could therefore have led to better nutritional conditions for some krill predators like the Antarctic minke whale. Although there is no direct observational evidence of improved nutritional conditions at that time, it is known that better nutritional conditions in whales may be reflected in a decrease in the age at sexual maturity. This was the case for Antarctic minke whales between approximately 1940 and 1970, which coincides with the period of depletion of some key krill-eating large whale species. This low age at sexual maturity led to an increase in the recruitment rate and total population size for minke whales over this period. The Antarctic minke whale perhaps rose to close to an increased carrying capacity resulting from a larger krill population by the 1970s, with the stock achieving this by stabilizing its age at sexual maturity at lower 7–8 years. The period of these demographic changes in Antarctic minke whales coincides with that characterized as 'favorable climate conditions and reduced competition for krill' and linked to penguin population changes [3].

6. Current ecosystem changes revealed through whale demography

Commercial whaling of humpback, Antarctic blue and fin whales in the Antarctic was suspended in 1963, 1964 and 1976, respectively. As an effect of these conservation measures, the abundance of these species have increased in recent decades, i.e. there have been recoveries from past commercial harvesting. The speed of recovery varies among species and populations. The increases in abundance of the recovering species will have effects on the ecosystem as a whole. In this section the current information on abundance and abundance trends of baleen whale species is examined, as well as some aspects of current nutritional conditions, feeding ecology and biological parameters of the Antarctic minke whale, a species which benefited from the krill surplus during the last century.

6.1 Abundance trend of baleen whales

Estimates of abundance of large whales in the Antarctic are based on systematic sighting surveys in open sea carried out under sampling methods advocated in DISTANCE [20], and the guidelines for surveys and analyses agreed by the IWC Scientific Committee (SC) [21]. Dedicated sighting surveys were carried out in the Indo-Pacific sector of the Antarctic under the JARPA/JARPAII programs using the IWC SC-agreed methodology and guidelines. Overall, the IWC SC carried out three circumpolar sighting surveys under the IDCR/SOWER programs (CPI, CPII and CPIII).

6.1.1 Humpback whales

Abundance and abundance trend estimates based on JARPA and JARPAII focused mainly in Areas IV (Population D) and V (Population E) (**Figure 1**). In Area IV the abundance was estimated in 29,067 whales (CV = 0.255) based on sighting data collected in 2007/08; in Area V the abundance was estimated in 13,894 whales (CV = 0.338) based on sighting data collected in 2008/09 [22].

Figure 6 shows the abundance trend of Populations D and E based on JARPA and JARPAII sighting data. For comparison purposes, the figure includes data from the IDCR/SOWER programs [23]. The figure shows a clear increasing trend, which is consistent for the JARPA/JARPAII and IDCR/SOWER survey data. Annual rate of increase was estimated at 13.6% (95% CI = 8.4–18.7%) and 14.5% (95% CI = 7.6–21.5%) for Areas IV and V, respectively, which were statistically greater than zero [22].

6.1.2 Antarctic minke whale

Abundance estimates and abundance trends of Antarctic minke whale for the Eastern Indian Ocean population (Area IV) and Western South Pacific population (Area V) have been conducted based on sighting surveys under JARPA [24]. Abundance estimates for the Eastern Indian Ocean population ranged from 16,562 (CV = 0.542) in 1997/98 to 44,945 (CV = 0.338) in 1999/00. Estimates for the Western South Pacific Ocean population ranged from 74,144 (CV = 0.329) in 2004/05 to 151,828 (CV = 0.322) in 2002/03.

Estimates of the annual rates of increase in abundance were 1.8% (95% CI: −2.5%, 6.0%) for the Eastern Indian Ocean population and 1.9% (95% CI: −3.0%, 6.9%) for the Western South Pacific population, which were not statistically greater than zero (**Figure** 7) [24].

The estimates based on IDCR/SOWER data were 55,237 (CV: 0.17) in Circumpolar II (CPII) and 59,677 (CV: 0.34) in circumpolar III (CPIII) for the Eastern Indian Ocean population (Area IV). For the Western South Pacific population (Area V) were 300,214 (CV: 0.13) in CPII and 183,915 (CV: 0.11) in CPIII.

6.1.3 Fin whales

For the purpose of the abundance estimates based on JARPA and JARPAII surveys, south of 60°S, the whole area was divided into a western area (Areas IIIE+IV) and eastern area (Areas V + VIW). For the western area the abundance was

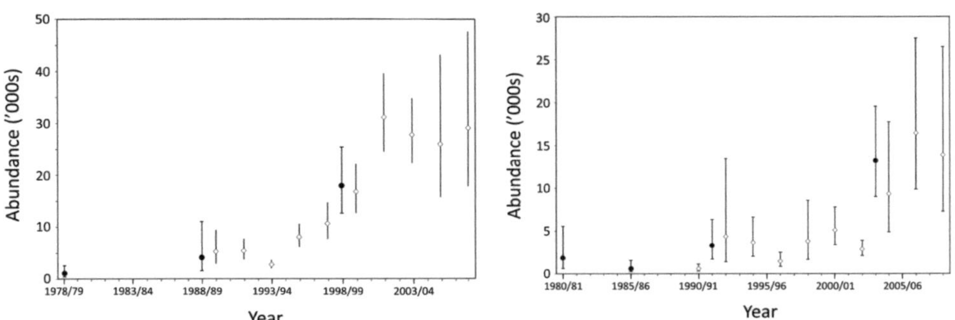

Figure 6.
Annual abundance trend for humpback whales in Areas IV (Population D) (left) and V (Population E) (right), south of 60°S. Estimates were based on sighting data collected by JARPA and JARPAII surveys primarily during January to February. Estimates from the IDCR-SOWER surveys [23] are shown for comparative purposes (filled circles). Vertical lines show 95% confidence intervals (modified from [22]).

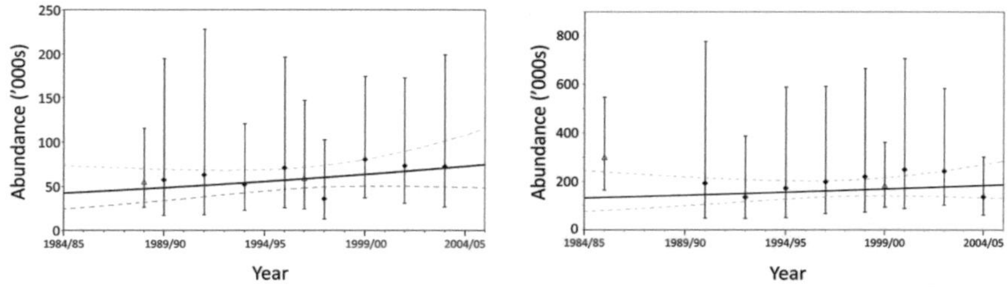

Figure 7.
Annual abundance trend for Antarctic minke whales of the Eastern Indian Ocean population (Area IV) (left) and Western South Pacific population (Area V) (right), together with their 95% CIs, based on sighting data from JARPA. The IDCR/SOWER estimates are shown for comparison (open triangles). The dashed curves indicate the 95% CIs for the exponential model applied to the JARPA estimates (modified from [24]).

estimated as 3,087 (CV = 0.191) in 1995/96, and 2,610 (CV = 0.285) in 2007/08. For the eastern area the abundance was estimated as 1,879 (CV = 0.226) in 1996/97, and 14,981 (CV = 0.298) in 2008/09. For the western area the increasing trend between 1995/96 and 2007/08 seasons was estimated at 8.9% (95%CI: -0.145%, 32.4%), while the trend in the eastern area between 1996/97 and 2008/09 was estimated at 12.0% (95%CI: 2.6%, 21.5%). The estimate for the eastern area was statistically greater than zero [25]. It should be noted that the JARPA/JARPAII surveys do not cover the latitudinal sector 50°-60°S, where fin whales distribute in large numbers in summer.

6.1.4 Antarctic blue whales

There is limited information on stock structure of Antarctic blue whales. Abundance of this species for the Indo-Pacific sector of the Antarctic (35°E-145°W), south of 60°S was 664 (CV = 0.328) in 2005/06 and 2006/07 seasons. The abundance was estimated at 1,223 whales (CV = 0.345) in the 2007/08 and 2008/09 seasons. The abundance trend was estimated at 8.2% (95% CI: 3.9%, 12.5%) between 1995/96 and 2008/09, which was not statistically greater than zero [25].

For most of the populations of these whale species were over-exploited in the past, but there is now scientific evidence of their recovery, although the speed of recovery is different among species and populations. The populations of Antarctic minke whale appear to be stable in recent years.

6.2 Changes in the distribution pattern of baleen whales

Substantial increases in the abundance of some species could have an implication on their pattern of distribution. Antarctic humpback whales from population D (Area IV) have increased substantially over recent decades, while the abundance of the Eastern Indian Ocean Population of Antarctic minke whale (Area IV) has been rather stable since the 1990s. We might expect some changes in the pattern of distribution of these two species. The spatial distribution of Antarctic minke and humpback whales was examined in the Indian sector (Area IV) based on JARPA/JARPAII sighting data for three periods: early (1989/1990, 1991/1992 and 1993/1994), middle (1995/1996, 1997/1998 and 1999/2000) and late (2001/2002, 2003/2004 and 2005/2006) [26]. Spatial distribution was estimated using generalized additive models (GAM). Presence or absence of whales was used as the response variable while seafloor depth, distance from shelf break and longitude were used as explanatory variables.

Mean probabilities of occurrence of Antarctic minke whales in the survey area in early, middle and late periods were 0.41, 0.46 and 0.41, while those of humpback whales were 0.14, 0.35 and 0.46. Occupied area indices (probabilities of occurrence of Antarctic minke whales less probabilities of occurrence of humpback whales) were also calculated. If the index is 1, only Antarctic minke whales were present in a grid cell, while only humpback whales were present if the index is −1. If the index is 0, probabilities of the presence of Antarctic minke whales and humpback whales in a grid cell were identical. Mean occupied area indices in early, middle and late periods were 0.28, 0.11 and − 0.07, respectively. The authors [26] concluded that the spatial distribution of humpback whales expanded to the south during the period investigated, while that of Antarctic minke whales remained stable. A summary of their results is presented in **Figure 8**.

The analyses were conducted based on sighting data obtained in the open sea. It should be mentioned that in the most recent period, Antarctic minke whales have also been observed frequently in polynias within the pack-ice [27], reflecting perhaps a response of this species to the geographical expansion of humpback whales to the south.

6.3 Changes in energy storage and stomach content weight

Nutritional condition in Antarctic minke whales has been investigated through different indices: blubber thickness under the assumption that the amount of lipids increase with the thickness of the blubber, girth and total fat. These data have been collected for more than 25 years during the JARPA and JARPAII surveys in the Indo-Pacific sector of the Antarctic (Areas IV and V). Regression analyses has shown that blubber thickness, girth and fat weight of sexually mature whales have been decreasing for nearly two decades [28]. The decrease per year was estimated at approximately 0.02 cm for mid-lateral blubber thickness and 17 kg for fat weight, corresponding to 9% for both measurements over the 18-year period (**Figure 9**).

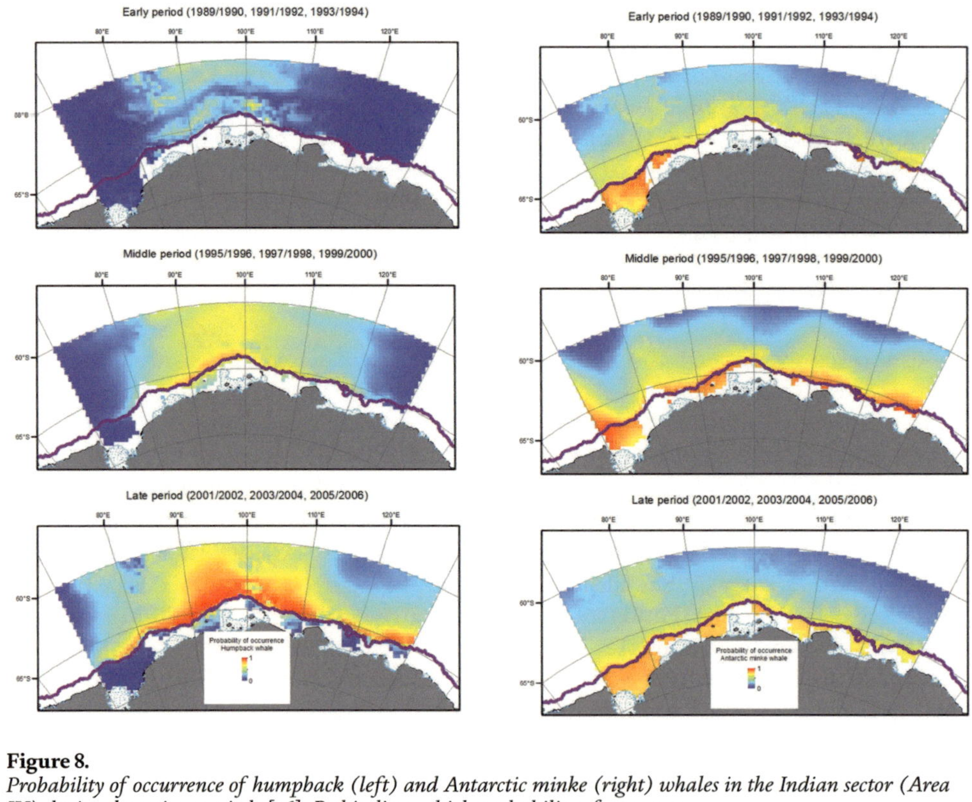

Figure 8.
Probability of occurrence of humpback (left) and Antarctic minke (right) whales in the Indian sector (Area IV) during three time periods [26]. Red indicates high probability of occurrence.

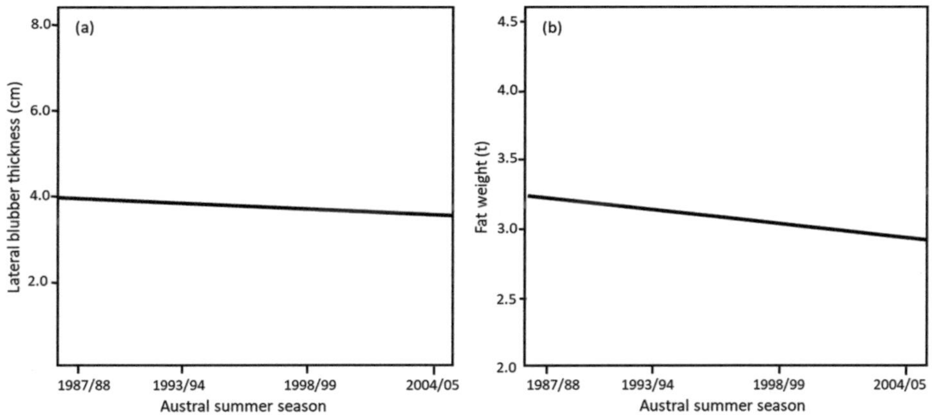

Figure 9.
Yearly trends in blubber thickness (a) and fat weight (b) for Antarctic minke whale in the Indo-Pacific sector of the Antarctic (modified from [28]).

Another study has reported the results of an analysis of temporal trend in stomach content weight in the Antarctic minke whale based on JARPA/JARPAII surveys [29]. A linear mixed-effects analysis showed a 31% (95% CI: 12.6–45.3%) decrease in the weight of stomach contents over the 20 years since 1990/91. A similar pattern of decrease was found for both males and females, except in the case of females sampled at higher latitudes in the Ross Sea (**Figure 1**). These results are consistent with the decline in energy storage reported above. Humpback whales are not found in the Ross Sea, where both Antarctic krill and ice krill (*E. crystallorophias*) are available, and where the authors [29] found no change in prey abundance for Antarctic minke whales.

The studies summarized above suggested a decrease in the abundance of krill for Antarctic minke whales, except in the Ross Sea.

6.4 Biological parameters

As indicated above, the ASM has remained stable at 7–8 years until the 1998 cohort (at least). The proportion of pregnant animals among mature females (PPF) for the Antarctic minke whale has been examined based on JARPA/JARPA surveys conducted between 1987/88 and 2010/11 [30]. The PPF for the Eastern Indian Ocean and Western South Pacific populations was high: 0.932 and 0.904, respectively, using data from all years combined. Linear regression analyses of the PPF over the years showed no significant temporal trend.

6.5 Interpretation of results

The results of this section can be summarized as following:

- The abundance of the once over-exploited large whale species such as the Antarctic blue, fin and humpback whales, has been increasing since the 1980's (at least). In particular, the increasing trends of Populations D and E of humpback whale and that of a population of fin whales in the eastern part of the research area were statistically significantly greater than zero.

- Since about 1990, the abundance of two populations of Antarctic minke whale has been rather stable as revealed by the SCCA analyses and abundance estimated by sighting surveys.

- Humpback whales from Population D (Area IV) have expanded their geographical distribution to the south in recent years, while the distribution of the Antarctic minke whales of the Eastern Indian Ocean population (Area IV) has remained stable in the open sea. Larger number of Antarctic minke whales are distributed in polynias within the pack-ice in recent years.

- The nutritional conditions of Antarctic minke whales has been deteriorating as evidenced by a decreasing trend of several nutritional indices.

- The ASM of Antarctic minke whale has remained low and stable around 7–8 years old until the 1998 cohort, and the PPF of Antarctic minke whale has remained high (over 0.9), and stable.

Deteriorating nutritional conditions of the Antarctic minke whales suggest less food available for this species in recent years. Less availability/abundance of Antarctic krill could be a response to environmental changes (e.g. global warming) and/or competition for food with the increasing large whale species such as Antarctic blue, fin and humpback whales. No evidence of global warming exists for the Indo-Pacific sector of the Antarctic [31], so competition for food could be a more plausible interpretation for the results of nutritional deterioration in the Antarctic minke whale. This reflects the end of the period of a krill surplus hypothesised during the past century. Nutritional deterioration as a consequence of competition is not entirely consistent with the low ASM and high APR of the Antarctic minke whale over more recent years. Under the competition hypothesis, this could be the result of a temporal phenomenon in that the response of ASM and APR to environmental changes producing a nutritional deterioration may be subject to time lags.

Direct competition occurs when two predators are present in the same area as a prey species, and may interfere with each other's access to the prey. Indirect competition may occur when two predators occur in different parts of the area of prey, but because the prey's production is limited, consumption by the one predator limits the production available for the other, and *vice versa* [32]. To investigate the plausibility of the competition hypothesis, estimates of krill biomass trends in the research area are required. There is some partial information on krill abundance based on dedicated krill surveys in the past, but the information is scattered and needs to be combined with that from new surveys in a comprehensive and consistent way so that time series data can be obtained. The period of these demographic changes in the Antarctic minke whales coincides with the post whaling era (1970's-) which has been characterized as 'unfavorable climate conditions and increasing competition for krill' [3].

Antarctic minke whales could also be using alternative feeding areas (e.g. polynias within the pack-ice) in response to the increase in abundance and geographical expansion of these other large whale species. This could provide an alternative explanation from sighting surveys and population models of a decrease and then re-stabilization of Antarctic minke whales abundance in open areas since the 1970's.

The increase of the Adelie penguins (*Pygoscelis adeliae*) in East Antarctic in recent decades [33] seems immediately not to be consistent with the competition hypothesis. The authors of the Adelie penguin study provided two explanations for the increase of this species: (i) harvesting of baleen whales, krill and fish across East Antarctic waters through the 20th century could have reduced competition between Adelie penguins and other predators for food, improving prey availability, and (ii) a proposed reduction in sea-ice extent in the mid-20 century may also have benefited Adelie penguins by enabling better access to the ocean for foraging. Since recovery of krill-eating large baleen whales has been reported since the 1980's, it is suggested that their explanation of environmental factors for the demographic changes of Adelie penguins since the

1980's is more plausible. Perhaps environmental changes have stronger effects on land-based predators such as the penguins than on sea-based predators such as whales.

7. Conclusions

This review of the scientific evidence for ecosystem changes in the Indo-Pacific sector of the Antarctic has highlighted the importance of long-time monitoring research programs focused on the collection of biological data for krill predators (both land-based and sea-based predators) as well as data on sea ice cover and environmental variables. The hypothesis proposed for the recent demographic changes found in whales is competition. To investigate this hypothesis further, estimates of krill abundance as well additional data collection of the predators and improved modelling analyses are required (see [34]). Also, oceanographic data obtained for more than 30 years in the Indo-Pacific sector of the Antarctic by JARPA/JARPAII surveys should be analysed to explore further the possible effect of global warming on the pattern of demographic changes found in whales.

In this context, Japan has started a new non-lethal research program to continue studying whales and the environment in the Indo-Pacific sector of the Antarctic. The program called JASS-A (Japanese Abundance and Stock structure Survey in the Antarctic) started in the austral summer of 2019/20, and will continue for at least eight years. It will conduct systematic sighting surveys for abundance estimates, biopsy sampling for genetic analysis of population structure, oceanographic and marine debris surveys, satellite tracking and photo-identification for studies on stock structure, distribution and movement of large whales, and Unmanned Aerial Vehicle (UAV) to observe whales outside of the main survey area. The analyses of the data to be collected will assist to examine the plausibility of the hypothesis proposed in this study further, in particular the observation that Antarctic minke whales have been moving into polynias within the pack-ice in recent years.

Acknowledgements

The authors thank the Fisheries Agency, Government of Japan for permits and funding for the whale surveys conducted in the Antarctic under the JARPA and JARPAII. They also thank the crew and researchers that participated in the surveys under the JARPA, JARPAII and IDCR/SOWER for collecting data and samples used in the analyses and studies summarized in this chapter.

Conflict of interest

The authors declare that there is no conflict of interest.

Author details

Yoshihiro Fujise and Luis A. Pastene*
Institute of Cetacean Research, Tokyo, Japan

*Address all correspondence to: pastene@cetacean.jp

References

[1] Laws RM. Seals and whales of the Southern Ocean. Philosophical Transactions of the Royal Society of London B. 1977;**279**:81-96

[2] Laws RM. The Ecology of the Southern Ocean. American Scientist. 1985;**73**:26-40

[3] Trivelpiece WZ, Hinke JT, Miller AK, Reiss CS, Trivelpiece SG, Watters GM. Variability in krill biomass links harvesting and climate warming to penguin population changes in Antarctica. PNAS. 2011;**108**(18):7625-7628

[4] Fraser WR, Trivelpiece WZ, Ainley DG, Trivelpiece SG. Increases in Antarctic penguin populations: reduced competition with whales or a loss of sea ice due to environmental warming? Polar Biology. *1992*;**11**:525-531

[5] Vaughan DG, Marshall GJ, Connolley WM, Parkinson CL, Mulvaney R, Hodgson DA, et al. Recent rapid regional climate warming on the Antarctic Peninsula. Climate Change. 2003;**60**:243-274

[6] Nicol S. Krill, currents, and sea ice: *Euphausia superba* and its changing environment. Bioscience. 2006;**56**(2):111-120

[7] International Whaling Commission. Report of the Workshop on the Comprehensive Assessment of Southern Hemisphere Humpback Whales. J. Cetacean Res. Manage. 2011;(special issue)**3**:1-50

[8] Kanda N, Goto M, Pastene LA. Stock structure of humpback whales in the Antarctic feeding grounds as revealed by microsatellite DNA data. Paper SC/F14/J31 presented to the Expert Workshop to Review the Japanese JARPAII Special Permit Research Programme, Tokyo, February 2014. 5pp. [Available from the IWC Secretariat]

[9] International Whaling Commission. Report of the Intersessional Workshop to Review Data and Results from Special Permit Research on Minke Whales in the Antarctic, Tokyo, 4-8 December 2006. J. Cetacean Res. Manage. 2008;**10**:411-445

[10] Pastene LA, Goto M. Genetic characterization and population genetic structure of the Antarctic minke whale *Balaenoptera bonaerensis* in the Indo-Pacific region of the Southern Ocean. Fisheries Science. 2016;**82**:873-886

[11] Kitakado T, Schweder T, Kanda N, Pastene LA, Walloe L. 2014. Dynamic population segregation by genetics and morphometrics in Antarctic minke whales. Paper SC/F14/J29 presented to the Expert Workshop to Review the Japanese JARPAII Special Permit Research Programme, Tokyo, February 2014. 20pp. [Available from the IWC Secretariat]

[12] International Whaling Commission. Report of the Expert Workshop to Review the Japanese JARPAII Special Permit Research Programme. J. Cetacean Res. Manage. 2015;**16**:369-409

[13] Lockyer C. The age at sexual transition of the southern fin whales (*Balaenoptera physalus*) using annual layer counts in the earplug. Journal du Conseil International pour l'Exploration de la Mer. 1972;**34**:276-294

[14] Thomson RB, Butterworth DS, Kato H. 1999. Has the age at transition phase of southern hemisphere minke whales declined over recent decades? Mar. Mamm. Sci. 1999;**15**(3):661-682

[15] Bando T, Kishiro T, Kato H. Yearly trend in the age at sexual maturity of Antarctic minke whales examined by transition phase in earplugs collected during JARPA and JARPAII surveys. Paper SC/F14/J8 presented to the Expert Workshop to Review

the Japanese JARPAII Special Permit Research Programme, Tokyo, February 2014. 10pp. [Available from the IWC Secretariat]

[16] Fournier D, Archibald CP. A general theory for analyzing catch at age data. Canadian Journal of Fisheries and Aquatic Sciences. 1982;**39**:1195-1207

[17] Punt AE. A summary history of the application of statistical catch-at-age analysis (SCAA) to Antarctic minke whales. J. Cetacean Res. Manage. 2014;**14**:81-92

[18] Punt AE, Hakamada T, Bando T, Kitakado T. 2014. Assessment of Antarctic minke whales using statistical catch-at-age analysis (SCAA). J. Cetacean Res. Manage. 2014;**14**:93-116

[19] Allen KR. Conservation and Management of Whales. Washington and Butterworths, London: Washington Sea Grand Publication; 1980 107p

[20] Buckland ST, Rexstad EA, Marques TA, Oedekoven CS. Distance Sampling: Methods and Applications. Springer International Publishing; 2015. 277p

[21] International Whaling Commission 2012. Requirements and Guidelines for Conducting Surveys and Analysing Data within the Revised Management Scheme. J. Cetacean Res. Manage. 2012;**13**:509-517

[22] Hakamada T, Matsuoka K. Estimates of abundance and abundance trend of the humpback whale in Areas IIIE-VIW, south of 60°S, based on JARPA and JARPAII sighting data (1989/90-2008/09). Paper SC/F14/J04 presented to the Expert Workshop to Review the Japanese JARPAII Special Permit Research Programme, Tokyo, February 2014. 36pp. [Available from the IWC Secretariat]

[23] Branch TA. Humpback whale abundance south of 60°S from three complete circumpolar sets of surveys. J. Cetacean Res. Manage. 2011;(special issue) **3**:53-69

[24] Hakamada T, Matsuoka K, Nishiwaki S, Kitakado T. Abundance estimates and trends for Antarctic minke whales (*Balaenoptera bonaerensis*) in Antarctic Areas IV and V based on JARPA sighting data. J. Cetacean Res. Manage. 2013;**13**(2):123-151

[25] Matsuoka K, Hakamada T. Estimates of abundance and abundance trend of the blue, fin and southern right whales in the Antarctic Areas IIIE-VIW, south of 60°S, based on JARPA and JARPAII sighting data (1989/90-2008/09). Paper SC/F14/J05 presented to the Expert Workshop to Review the Japanese JARPAII Special Permit Research Programme, Tokyo, February 2014. 36pp. [Available from the IWC Secretariat]

[26] Murase H, Matsuoka K, Hakamada T, Kitakado T. Preliminary analysis of changes in spatial distribution of Antarctic minke and humpback whales in Area IV during the period of JARPA and JARPAII from 1989 to 2006. Paper SC/F14/J18 presented to the Expert Workshop to Review the Japanese JARPAII Special Permit Research Programme, Tokyo, February 2014. 17pp. [Available from the IWC Secretariat]

[27] Ishikawa H, Goto M, Ogawa T, Bando T, Kiwada H, Isoda T, Kumagai S, Mori M, Tsunekawa M, Ohsawa T, Fukutome K, Koyanagi T, Kandabashi S, Kawabe S, Sotomura N, Matsukura R, Kato K, Matsumoto A, Nakai K, Hasegawa M, Mori T, Yoshioka S, Yoshida T. Cruise report of the Second Phase of the Japanese Whale Research Program under Special Permit in the Antarctic (JARPAII) in 2007/2008. Paper SC/60/O4 presented to the IWC Scientific Committee, Anchorage, June 2007. 19pp[Available from the IWC Secretariat]

[28] Konishi K, Tamura T, Zenitani R, Bando T, Kato H, Walløe L. Decline in energy storage in the Antarctic minke whale (*Balaenoptera bonaerensis*) in the Southern Ocean. Polar Biology. 2008;**31**:1509-1520

[29] Konishi K, Hakamada T, Kiwada H, Kitakado T, Walløe L. Decline in stomach contents in the Antarctic minke whale (*Balaenoptera bonaerensis*) in the Southern Ocean. Polar Biology. 2014;**37**:205-215

[30] Bando T, Hakamada T. Yearly trend in the proportion of pregnant animals among mature female Antarctic minke whales in the JARPA and JARPAII period. Paper SC/F14/J9 presented to the Expert Workshop to Review the Japanese JARPAII Special Permit Research Programme, Tokyo, February 2014. 6pp. [Available from the IWC Secretariat]

[31] Watanabe T, Okazaki M, Matsuoka, K. Results of oceanographic analyses conducted under JARPA and JARPAII and possible evidence of environmental changes. Paper SC/F14/J20 presented to the Expert Workshop to Review the Japanese JARPAII Special Permit Research Programme, Tokyo, February 2014. 10pp. [Available from the IWC Secretariat]

[32] International Whaling Commission. Report of the Scientific Committee. J. Cetacean Res. Manage. 2015;**16**:1-87

[33] Southwell C, Emmerson L, McKinlay J, Newbery K, Takahashi A, Kato A, et al. Spatially extensive standardized surveys reveal widespread, multi-decadal increase in East Antarctic Adelie penguin populations. PLoS One. 2015 http://dx.doi.org/10.1371/journal.pone.0139877

[34] Murase H, Palka D. Punt A, Pastene LA, Kitakado T, Matsuoka K, et al. Review of the assessment of two stocks of Antarctic minke whales (Eastern Indian Ocean and western South Pacific). J. Cetacean Res. Manage. 2020;**21**:95-122

Chapter 6

Risks of Glaciers Lakes Outburst Flood along China Pakistan Economic Corridor

Muhammad Saifullah, Shiyin Liu, Muhammad Adnan, Muhammad Ashraf, Muhammad Zaman, Sarfraz Hashim and Sher Muhammad

Abstract

The China-Pakistan Economic Corridor (CPEC) passes through the Hunza River basin of Pakistan. The current study investigates the creation and effects of end moraine, supra-glacial, and barrier lakes by field visits and remote sensing techniques along the CPEC in the Hunza River basin. The surging and moraine type glaciers are considered the most dangerous type of glaciers that cause Glacial Lake Outburst Floods (GLOFs) in the study basin. It can be concluded from the 40 years observations of Karakoram glaciers that surge-type and non-surge-type glaciers are not significantly different with respect to mass change. The recurrent surging of Khurdopin Glacier resulted in the creation of Khurdopin Glacial Lake in the Shimshal valley of the Hunza River basin. Such glacial lakes offer main sources of freshwater; however, when their dams are suddenly breached and water drained, catastrophic GLOFs appear and pose a great threat to people and infrastructure in downstream areas. This situation calls for an in-depth study on GLOF risks along the CPEC route and incorporation of GLOF for future policy formulation in the country for the CPEC project so that the government may take serious action for prevention, response to GLOFs, and rehabilitation and reconstruction of the areas.

Keywords: glaciers, GLOF, CPEC, climate change, lakes

1. Introduction

The high altitudes of the China Pakistan Economic Corridor (CPEC) region encompass glaciers as frozen reserves of water which act as an important natural resource by supplying fresh water to millions of people living the mountainous and downstream areas. The water released from the glaciers acts as a perennial source for most of the Himalayan Rivers [1]. The rivers and streams originating from these glaciers not only serve as power generation from hydroelectric power plants but also irrigate the agricultural lands in the command area during summer and also provide water for industrial purposes. Like this, these glaciers control the socio-economic activity in this part of the CPEC region. Therefore, the meltwater from the snow and glaciered region is of high importance for the runoff in the Indus River [2]. There is a lack of exact facts and figures about the exact contribution of flow in the region due to rugged mountains and the limited data availability because there is a highly different precipitation rate due to the settings and steep topography. Similarly, there are extremely different ablation rates due to aspect and variable debris cover on the glaciers. However, an extensive field study was conducted by the Snow and Ice Hydrology Project (SIHP) of Water and Power Development Authority (WAPDA) with the collaboration of Canadian University during the 1980s. It was reported that the maximum precipitation occurs at the elevation of 4500–5100 m.a.s.l. Moreover, Hewitt [3] reported that about 80% of flows in the Upper Indus River derives from the glaciered region above 3500 m.a.s.l. Recently, researchers have tried to assess the spatial distribution of precipitation by inversely inference of precipitation for glacier mass balance [4] snow cover variation [1, 5] using remote sensing data, Geographical Information System (GIS) techniques, and runoff modeling approaches. Adnan et al. [5] reported that the expansion in snow cover in the Hunza River could be attributed to the surge activities in the basin. Many researchers [6–10] have reported the surge events in the region. Studies on glaciers also indicate the slightly reduced mass balance in the region. However, most of the glaciers have gained the mass in the nourishment zone and loss in the ablation zone. These findings suggest the increase in the slopes of the glacier that could cause increased glacier velocities and the probable advance of glaciers in the future. However, in this chapter, we have presented the existence of glaciers and glacial lakes in the Hunza basin through which the CPEC route passes and is a heavily glaciered region. Many GLOF events and surges have been reported in the basin, especially, along the CPEC route. Moreover, glaciers are sensitive indicators of global climate change because they remain sensitive to global temperature conditions as specified by their continuous retreat which has been witnessed in many parts of the world including the Hindu-Kush Karakoram Himalaya (HKH) region especially CPEC route [9, 11]. After the industrial revolution, the rapid glaciers melting and its associated retreating trend left a major concern to the scientists and managers in the region. The substantial glaciers melting not only decrease the rivers discharge in the long run but also bring the high sediment load which causes flash floods in downstream areas and has a direct posture on the life of hydropower projects and socio-economic consequences for the local people and those living in downstream areas.

1.1 Glacial Lake Outburst Floods (GLOFs)

Terminal moraines and glacial lakes have been exposed in these high mountains as a result of surging, substantial melting, and retreating of glaciers. GLOFs have become a matter of concern for social and economic stability in the river valleys because of the rapid addition of meltwater in those glacial lakes adjacent to receding valley glaciers and may lead to sudden breaching of the unstable debris dam. Thus, it is very important to understand the state and fate of these glaciers and lakes as well for long-term development and planning in this region [12, 13].

During the last half-century, a large number of glacial lakes development have been witnessed in the CPEC region, and at the same time, several GLOFs have been reported, especially, in the eastern part of the CPEC region. Probably, remote glacial lakes are under risk and they may impact the downstream inhabitant as a result of GLOFs. These GLOFs may have devastating effects on the population as well as property and infrastructure [14–16].

The understanding that GLOFs can significantly surpass design floods of Hydropower Plant (HPP) under the threat of destruction or complete non-functionality is based on few case studies [17–19] but a clear picture of regional GLOF exposure remains vague. The previous glacial lake inventories to identify future GLOF sources have neglected their impacts on downstream areas [20]. However, hydrodynamic modeling may encounter such kinds of impacts but it requires a detailed analysis of high-resolution Digital Elevation Models (DEMs) and high computation facilities so that simulations of possible outbursts are existing for only a handful of lakes [12, 21, 22]. This research gap has been filled by linking a glacial lake inventory with both flood propagation and a dam-breach model to assess potential flood magnitudes for a sample of operating, planned, and currently built hydropower plant in the CPEC region [23–25].

Regional predictions of peak discharges as a result of GLOFs are difficult because of the unavailability of data on the geometry of glacial lakes and the moraine dams. The current field visit as well as different researchers provided an alternative solution to this problem and use the range of the simulated peak discharges at each HPP as a measure of uncertainty of GLOF exposure. For our flood propagation model, the volume and the resulting distributions of peak discharges for each lake have been used as the initial conditions during the field visit.

2. Hunza River basin

CPEC is in the interest of both the countries China and Pakistan; it will develop connectivity between west China and south China; and it is an integrated part of the "One Belt One Road" initiative policy. The establishments of China and Pakistan have agreed to complete the CPEC route from Kashgar (China) to Gwadar (Pakistan) by the end of 2030. The Chinese government is providing necessary support in terms of finance and logistics to build the infrastructure along the CPEC route. The CPEC is not only important for both the countries but also will prove beneficial for the surrounding countries. This project will strengthen the economic growth of Pakistan and it is the right initiative for both the countries. This corridor is considered a sign of peace, prosperity, and development. Even though this economic corridor is challenging but it will open new horizons of development in the future for both the countries. In the past, the lack of the right decision and insufficient opportunities have always remained a hurdle in the way of Pakistani peoples but CPEC will have transformational impacts on the state and the prosperity Pakistani nationals. The Chinese president Xi's visit in April 2015 and the announcement of $46 billion-plus for various CPEC projects drew the world's attention to this region because new development and growth of the economy will benefit both of the countries under the umbrella of this economic corridor. In the meeting called by the Prime Minister of Pakistan, all political parties have supported the CPEC and project and warmly welcomed the Chinese investment. This project will bring a revolution in the lives of over 3 billion people in this region. It will improve the strategic and economic location of Pakistan through trade and investment and exploration of mineral resources. Alternatively, according to China's perspective, this is a "flagship project" because it provides the shortest route to the Middle East, Africa, and Europe, which will boost up its economy. This corridor is passing through the Northern part of Pakistan.

The northern part of Pakistan is mostly consisting of a mountainous region, which is rich in glaciers and glacial lakes. The source of water in the river is glacier melting and rainfall. The population living downstream is under high risk due to the melting of glaciers and GLOFs [26–28]. International Center for Integrated Mountain Development (ICIMOD's) published an inventory in 2005, which comprised of 2420 glacial lakes in 10 major river basins of Hindukush Karakoram Himalayan Region of Pakistan [29]. The different river basins have glacial lakes such as Gilgit (614), Indus (574), Swat (255) Shingo (238), and Hunza (110). The Gilgit River basin comprised of 614 glacial lakes and 380 lakes out of 614 were identified

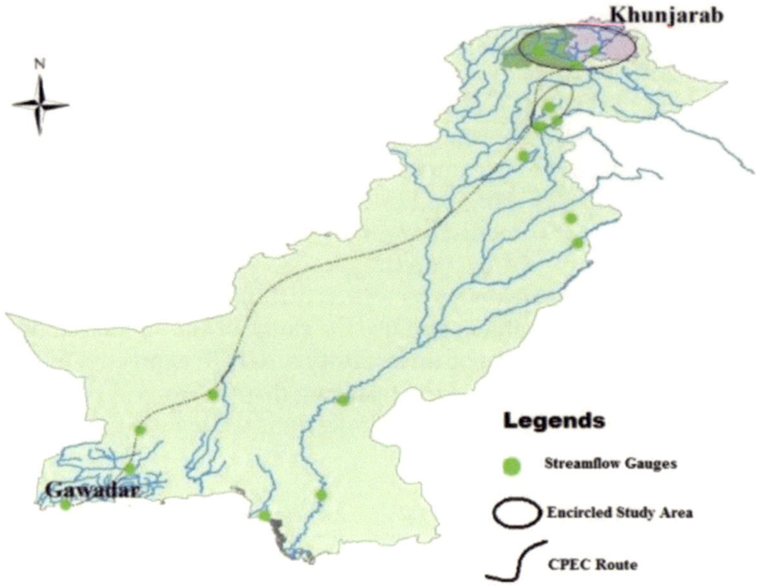

Figure 1.
Study area of China Pakistan economic corridor.

as major lakes, which contribute 62% of the total lakes. These major lakes form 93% of the lake area of the basin. New glacial lakes also formed due to glaciers thinning and retreating of this region. These lakes are categorized according to risk, 52 glacial lakes identified in this region. Passu lake has experienced historical GLOF events, which lie in the Hunza River basin [6]. The location of Passu lake is 38 km away and directed to the East-West of Passu glacier in the HKH region (**Figure 1**). The climate of Hunza is moderate, which have minimum and maximum temperature of 16 and 35.9°C, respectively. The annual average rainfall in this region is 136.2 mm with a minimum (2.1 mm) and maximum (283.2 mm) in November and April, respectively [30].

3. Materials and methods

3.1 Glaciers along CPEC route

The CPEC route passes through the Hunza River basin which is a glaciated region of Gilgit Baltistan. The route starts from China to Pakistan through the Sost border which lies in the Hunza River basin. Approximately, 28% of the Hunza basin area is covered by glaciers, and Passu, Batura, and Ghulkin are some known glaciers that exist along the CPEC route (**Table 1**), which have an established history of GLOF events. The Karakoram Highway and other roads in Shimshal and the Nagar River basins have been damaged many times because of GLOF events from the glaciers.

3.2 Assessment of GLOF using remote sensing data and GIS

Rugged mountain conditions make it difficult to investigate the glacial lakes for the whole region. However, the end moraine and the lakes in the blocked river valleys (e.g., Khurdopin glacial lake) were investigated with physical visits. However, the area's calculations and the causes of surging were made through remote sensing data and GIS

River basins	Glaciers			Glacial lakes		
	no.	Area (km²)	Ice reserve (km³)	No.	Area (km²)	Potential danger
Swat	233	223.55	12.221	255	15.86	2
Chitral	542	1903.67	258.817	187	9.36	1
Gilgit	585	968.1	83.345	614	39.17	8
Hunza	1050	4677.34	808.794	110	3.21	1
Shigar	194	2240.08	581.27	54	1.09	0
Shyok	372	3547.84	891.8	66	2.68	6
Indus	1098	688	46.38	574	26.06	15
Shingo	172	36.91	1.009	238	11.59	5
Astore	588	607.03	47.931	126	5.52	9
Jhelum	384	148.18	6.943	196	11.78	5

Table 1.
Summary of glaciers, glaciers lakes, and potentially dangerous lakes in CPEC region [31].

techniques. The Landsat satellite images for May and June 2017 were downloaded from the website; http://earthexplorer.usgs.gov/ to explore the formation of newly developed lakes due to river blockage caused by the surging of Khurdopin glacier. The formation of the newly developed lakes was identified from the visual interpretation of the images in the ArcGIS tool, whereas Temporal Geodetic mass balance is employed to compute the vertical changes in glaciers by using remote sensing data of that region. The potential GLOF lakes were identified base on the following criteria and physical conditions [24]:

The rise in water level in glacial lakes, which creates the condition to breach the lake.

The lakes form on the glacial surface, which produces the combined effect with a group of lakes. It became potentially dangerous lakes.

The valley lake also becomes the potential GLOF lake due to short distance from mother glaciers tongue and the size of lake also plays an important role.

A lake several times breaches and damages the downstream. These types of lakes filled again and breach again.

The physical conditions of the surrounding of the lake also play a vital role to identify the potential dangerous glacial lake.

There is still no standard to identify the potential glacial lakes. The above mention criteria and condition decide the potential dangerous glacial lakes.

3.3 Field investigations for GLOF events

Remote sensing and field investigations are two basic methodologies used to assess the GLOF events (**Table 2**) and their credible effects. The number of glacial lakes, their areas, and geodetic mass balance for surging glaciers have been estimated through remote sensing techniques; whereas, field investigations help to assess the severity of the GLOF event. Moreover, the possible disaster from the potential hazard lake can be investigated through field investigations in terms of barrier strength, discharge conditions, and depth of the lake.

Pakistan Snow and Ice Hydrology Project (PSIHP) of WAPDA in collaboration with a Canadian University studied the Khurdopin glacier for precipitation input. They also did field visit of Batura glacier and Passu lake through the collaboration of Joint Venture of National Science Foundation of China, under the umbrella of the institute of international Rivers and Eco-security, Yunnan University,

Year	Event date	Glacier	River	Influencing factors
1973	—	Batura	Hunza	—
1974	—	Batura	Hunza	—
1977	—	Balt Bare	Hunza	—
1978	September	Darkot/Barados	Gilgit	—
1994	July	Sosot/Gupis lake	Gilgit	—
1999	6 August	Khalti/Gupis	Gilgit	Moonsoon rainfall
2000	10 June	Shimshal	Hunza	High temperature
2000	27 July	Kand/Hushe	Indus	Moonsoon rainfall
2005	July	Sosot/Gupis lake	Gilgit	
2007	5 April	Ghulkin	Hunza	Western disturbance
2008	6 January	Passu	Hunza	Western disturbance
2008	2 April	Ghulkin	Hunza	Western disturbance
2008	22 May	Ghulkin	Hunza	Persistent rainfall
2008	24 May	Ghulkin	Hunza	Persistent rainfall
2008	14/15 June	Ghulkin	Hunza	Heat wave
2009	26 March	Ghulkin	Hunza	Western disturbance
2018	17 July	Barsuwat glacier	Immit	Heat wave
2019	23 June	Shishper	Hunza	High temperature
2020	29 May	Shishper	Hunza	High temperature

Table 2.
History of major GLOF events in CPEC region.

Kunming, China, and ICIMOD. The remote sensing and field observations analysis of Khurdopin glacier provides up to date evidence about glacier surge and its possible impacts on the downstream populations because of the newly developed lake.

4. Results and discussion

4.1 GLOF events in Hunza basin

Glacial lakes formation depends on the type of glacier, its slope, and geological settings. For example, a supraglacial lake creates on the surface of glaciers when its slope is less than 2%. Similarly, surging glaciers block the river valley and cause the formation of the lake. Moreover, moraine-dammed lakes develop due to the advancement or recession of valley glaciers. Overall, these GLOF events depend on the physical and topographical conditions and the nature of damming materials. The severity of damages increases as the elevation and the volume of the glacial lakes increases. The type of moraine-dammed and the surging glaciers are the most dangerous types that block the valleys and cause GLOF events in the basin. In this regard, the type of GLOF event for the glaciers in the Hunza Basin is also not the same for example; Passu glacier caused damage due to outbursts of the end-moraine dammed lake, supra-glacial lake outburst occurred from Ghulkin and Hispar glaciers, valley blocked by Khurdopin glacier. GLOF from these glaciers bring rocks and the mudflows in the glacial streams, for example, significant mudflows released from Batura glacier.

4.2 GLOF events associated with surging of glaciers

The surging activities of the glaciers in the Hunza Basin have also been interpreted by earlier studies [14, 32, 33] from a stable or slightly increasing trend of snow cover for the Hunza River basin [1, 25]. It has been reported recurrent surges for several glaciers of the Hunza Basin [6]. For example, Bolch et al. [20] and Quincy and Luckman [34] have comprehensively reported the surge history of Khurdopin glacier. The glaciated area of Khurdopin glacier is 115 km^2 which is situated in Shimshal River, a tributary of the Hunza River. The first surge has been reported in 1979 and the second surge event occurred in 1999 and both surge events occurred in the summer season. The latest surge was observed during the summer season of 2017 [35]. These events suggest the return period of the surge event for Khurdopin glacier is about 20 years. No significant change has been observed in the debris cover Hispar and Shimshal glaciers of the Hunza basin for the period of 1977–2014 [36]. It was determined that this might be due to balanced glacier budgets during this period. Type of glaciers and their areas are given in **Table 1**.

The surging of Khurdopin glacier has resulted in the formation of the medium-sized ice-dammed lake at an elevation of 3454 m a.s.l and it lies at latitude-longitude of 36°21′007″ N and 75°27′51.2″ in the Shimshal River valley of the Hunza basin. Khurdopin lake started to surge in the first week of May 2017 and it has been greatly expanded in terms of size and depth and it became vulnerable to breach as witnessed by the local people. Due to the short distance between the glacier and opposite hard mountain resulted in the rise of river bed and glacier as well as triggering the creation of Khurdopin lake. The velocity of the flow was reduced by the damming of water behind the barrier and this phenomenon also raised the river bed and blocked the river (**Figures 2** and **3**).

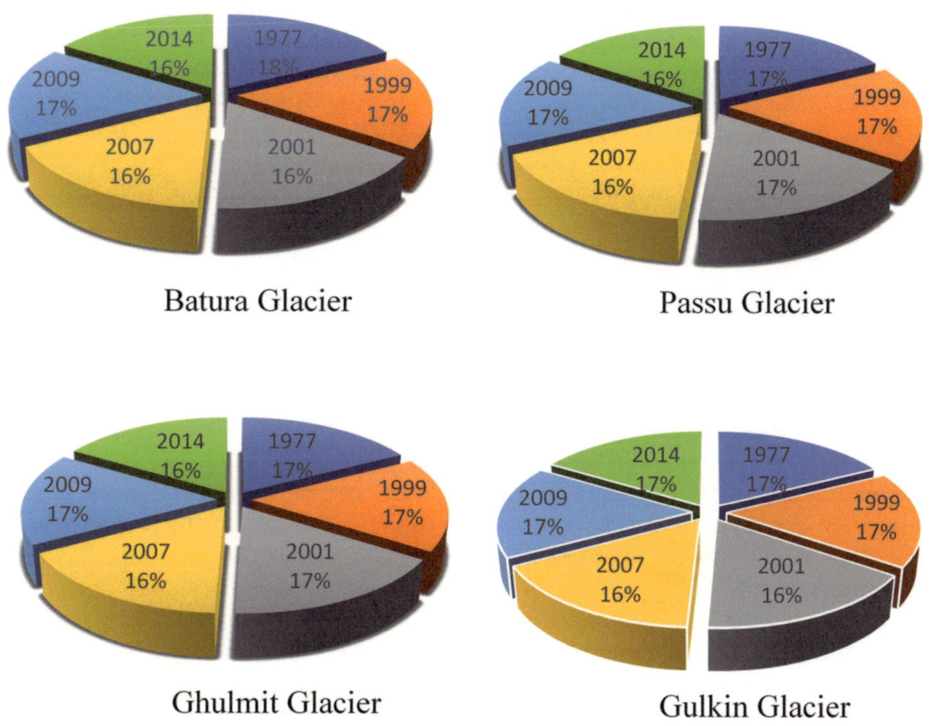

Figure 2.
Comparison of different glaciers area loss during the period of 1977–2014.

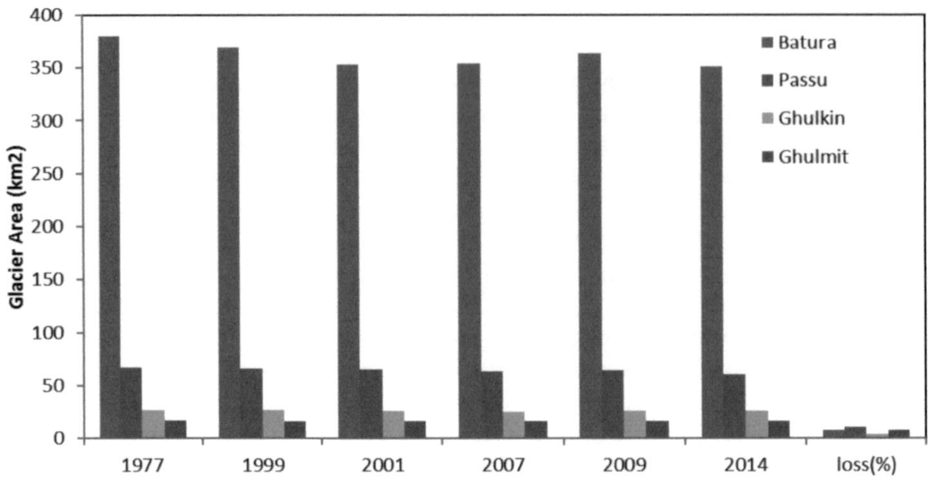

Figure 3.
Temporal variation in areas of different glaciers along CPEC.

The glacier areas of different glaciers were compared, which mostly thinned in terms of area or remain constant during the different periods. The area of Batura glacier was 380 km^2 in 1977 as compared to 351 km^2 in 2014, respectively. The loss of the area was 7.63% during this period. However, the glacier area loss was observed about 4.69% between 1999 and 2014. The glacier area was reduced to 369 km^2 till 1999 and lost 2.89% of the glacier area as compared to 1977. The loss in glacier area was increased by up to 7% in 2014 as compared to 1999. The increment of a 4% loss in glacier area was due to an increase in temperature. During 2001 and 2007, the loss of glacier area was consistent, but it was observed 4.2% in 2009 as compared to 1977. Passu glacier area loss observed 10% from 1977 to 2014. The loss of the area was less than 1% for 1999 as compared to 1977. The rate of loss of the glacier area increased in 2001 and reached up to 3% but this rate reached up to 7% in 2007. The lake formation also fluctuated during this period. The number of GLOF and historical events was also observed during this period. Ghulkin glacier also lost its area up to 4% due to global warming from 1977 to 2014. Earlier 2000, Ghulkin glacier lost the area less than 1.5%. After 2 years, the loss of the glacier area was 6%. From 2001 to 2007, the melting of glacier and loss of its area remained constant but the rate of glacier area loss was reduced up to 4% in 2009. Ghulmit glacier lost its area about less than 2% before 2000. The loss of the glacier area in 2001 was 4% but it increased up to 5% in 2007. This glacier area was reduced to 3% in 2009 as compared to 2007 (**Table 3**).

4.3 Drifting mechanism of glaciers

Due to differential movement of glaciers, crack or rift is produced in the glaciers having connected snout. A rift was formed along the left part of the Khurdopin glacier due to collapsing and crushing of glacier all together. The Yukshin Girdan glacier played a role of strong obstruction, produced massive frictional forces, and initiated the glacier ice to break down at the rift area into pieces of ice bergs that were seen floating on the newly formed lake. During this process, the Khurdopin glacier was forced by the Yukshin Girdan glacier to move toward the right side because of the unavailability of an obstacle to the opposite mountain. The elevation of the glacier at the rift area was about 3558 m a.s.l. The glacier was covered by a huge amount of debris during 2015, and after this surge, the debris began to falling

Sr.#	Glacier name	Glacier type	Area (km^2)
1	Batura	Debris cover	236
2	Passu	Debris free	51
3	Barpu	Surge type	90
4	Hispar	Surge type Debris cover	345
5	Yazghil	Debris cover	99
6	Khurdopin	Debris cover Surge type	115
7	Vijerab	Debris cover Surge type	113
Whole region			2868

Table 3.
List of selected glaciers and their types in Hunza basin [20].

and sliding into the crevasses and impacted the sub-glacial water movement and that may further result in the formulation of a new sub-glacial lake [25, 34].

The meltwater released from the Vijerab glacier was blocked by the Khurdopin glacier and it was muddy because of high sediment load coming from the glaciated area. A large amount of water was flowing through the newly developed crevasses due to the presence of barriers and ultimately outflowing through the cave type snout of Khurdopin glacier. The phenomenon of ice block falling in the stream of Vijerab glacier and on the supra-glacial lakes was also perceived. Because of surging effects, glacier deforms and results in the formation of huge crevasses. Three main types of crevasses developed in Khurdopin glacier are listed below: irregular crevasses, longitudinal crevasses, and transverse crevasses. Due to the demolishing action of glaciers, most of the crevasses were found to be irregular in shapes. Due to glacier advancement, large numbers of crevasses were developed all over the glacier. The width of the crevasses was variable at different locations and mostly, it ranged between 1 and 2.0 m [20, 35].

4.4 Geodetic mass balance of surging glaciers

The heterogeneous behavior and close-to balanced budgets are not a recent phenomenon in Hunza Valley (Karakoram). We have observed that the geodetic mass budgets computed from the 1973 KH-9 and 2009 ASTER DTMs are in covenant with the results of the individual periods 1973–1999 and 1999–2009 without indefinite radar penetration correction: since the 1970s, the glaciers in this region underwent slight and insignificant mass loss. Though, the differences may exist in individual glaciers for the two studied periods. Overall, we can confirm that the surge-type and non-surge-type glaciers are not significantly different with respect to mass change inferred from the 40 years observations of glaciers in the Karakoram (**Figure 4**).

One can easily assess the flood damages from recently developed lakes if it bursts based on watermarks of previous flood events. Besides, the bursting mechanism and the volume of flood events can give us an insight into the damages in the downstream areas. The water level in the lake can provide us information about the GLOF impacts on downstream infrastructure. The GLOF will have devastating impacts on the infrastructure including homes, lands, schools, etc. The settlement along the river site could be adversely affected as a result of GLOF (**Figure 5**).

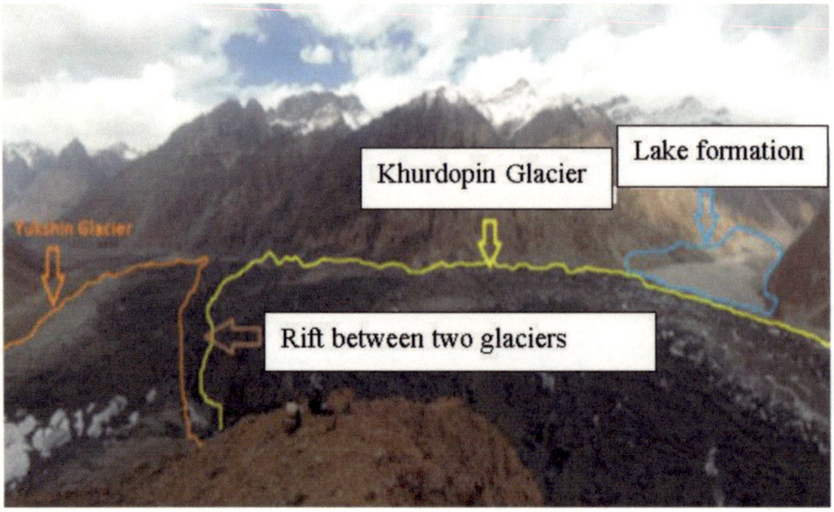

Figure 4.
Khurdopin Glacial Lake formation.

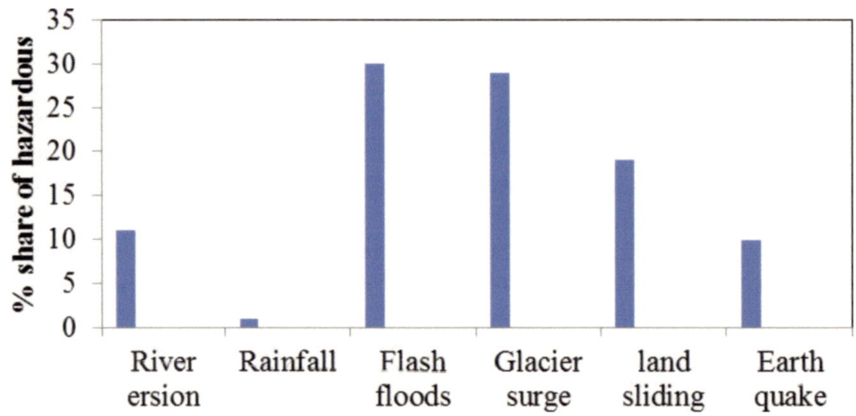

Figure 5.
Contribution of hazardous share in different glaciers along CPEC.

4.5 GLOF events associated with end-moraine dammed lakes

During 2001, the total numbers of lakes in Gilgit were 614. The total area of these lakes was 39.2 km^2. Out of 614 lakes, 380 were major and 8 were potential GLOF lakes. There were 110 glacial lakes in the Hunza basin, which have a total area of 3.2 km^2. Out of 47 glacial lakes, only one lake was potential GLOF, that is, Passu lake which has a potential threat to CPEC (**Figures 6** and 7). In the past, this lake had busted many times during the flooding period as given in **Table 2** and **Figures 6** and 7.

Amongst the end moraine dammed lakes of Hunza Basin, six were identified as potentially hazardous lakes for the CPEC route (**Figures 6** and 7). The major lakes were valley type and superglacial lakes. Passu lake was observed as a hazardous lake, which is near to Passu glacier. The surface area of this lake was 0.12 km^2, length 26 km, and thickness of about 173 m. The population of the Ghulkin, Hussaini, and Passu villages are 1133, 621, and 863 persons, respectively. During the July 2007 and April 2008, heavy flooding occurred, this damaged the CPEC route.

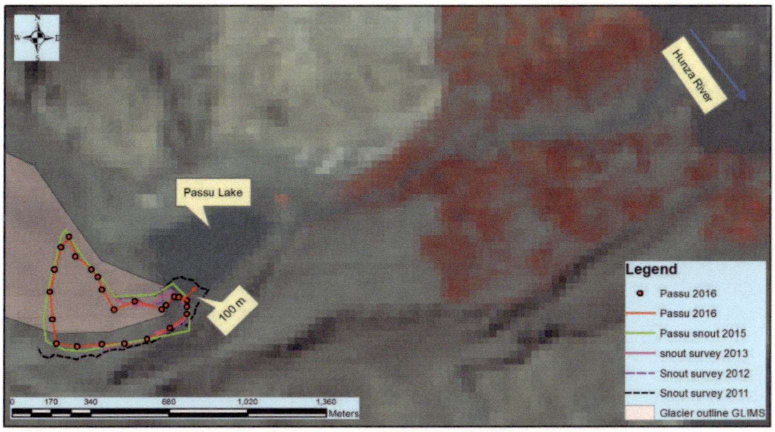

Figure 6.
Retreat of Passu glacier caused expansion in the glacial lake.

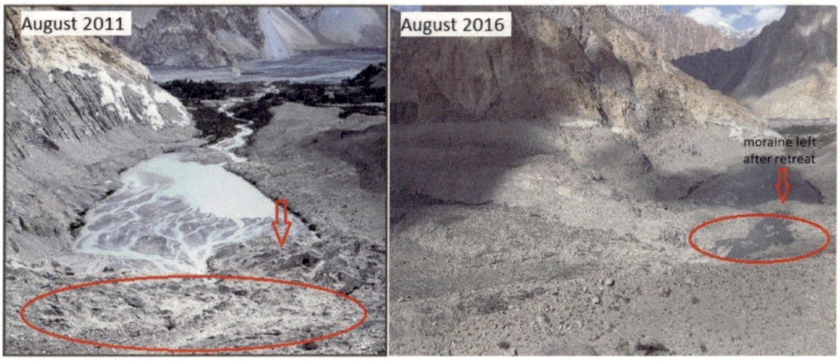

Figure 7.
View of the Passu glacier during 2011 and 2016 shows the expansion in lake area.

Basin	Number of lakes	Lake area	Major lakes	GLOF lakes
Gilgit	614	39.17	380	8
Hunza	110	3.22	47	1

Table 4.
Summary of glacial lakes in two major glaciered river basins along CPEC [20].

In July 2007, the lake was outburst with heavy flooding and another event was observed in April 2008. These events damaged the Karakoram Highway, hotels, and houses of the Passu village. This lake was breached several times in the past. Passu village is under high risk due to GLOF of this lake. There is a need to install the proper monitoring system to reduce the risks of glacier lakes outburst as well as local community protection in the downstream area (**Table 4**).

Although this lake is hazardous for the nearby communities of the Passu village and after the creation of a large land-slide dammed lake (Attabad lake) at Attabad during February 2010, the villages along Hunza River up to Gilgit and downstream became highly vulnerable of GLOF hazard. Early warning systems and proper monitoring can reduce the risk of damage to the CPEC route, infrastructures, and community. The different lakes of Gilgit have different rates of expansion such as 1.2, 0.4, and 0.3 ha/year during 2001 and 2013, respectively (**Figures 8–10**).

Figure 8.
Field observation of Passu glacier and lake.

Figure 9.
Impact of high flow of Batura glacier on CPEC bridge.

5. Conclusions

The rugged topography and remoteness of the study area is a hurdle in the way to study the detailed processes and reasons behind the creation of glacial lakes along the CPEC. However, remote sensing techniques along with field surveys helped us to study the remote lakes along the CPEC route in the Hunza River basin. It was inferred from the analysis that the glaciers and glacial lake dynamics are interconnected to each other. The glacial lake dynamics is a complex phenomenon along CPEC. Gradually, lake dynamics has increased the risks of vulnerability along CPEC. Many potential lakes in the study area have the capability to damage the infrastructures as well as routes. Several GLOF events have been reported, especially, in the eastern part of the CPEC region during the last 50 years. Recently, Shisper Glacier damaged

Figure 10.
(a) Shisper glacier and its recent impact on CPEC route; (b) view of Passu glacier and CPEC route alignment.

the CPEC route and blocked it (**Figure 10a**). In the same way, the repeated surging of Khurdopin and Passu glaciers also resulted in the formation of high altitude glacial lakes in the Hunza River basin. We found that the type of GLOF event for the glaciers in the Hunza River basin is not the same. The aerial expansion of these glacial lakes increased due to global warming. Several glaciers are retreating in this region; this retreating will result in reduced river flows which in turn will affect the available run-off for irrigation and power generation. Moreover, the history of Khurdopin glacier's surge events revealed that the thermal phenomenon causes these surges. Moreover, it was perceived that the Passu Glacial Lake expansion is due to the retreat of the glacier (**Figure 10b**). However, the structure of the glacier surface suggests that its advancement is due to an increase in slope. A maximum increase in lake area was observed below 3500 m elevation, exhibiting a situation favorable for water resource management. The climate and hydrodynamics also influence the glaciers and glacial lakes. The CPEC initiative will bring a revolution in the lives of over 3 billion people in this region. Being the shortest route to the Middle East, Africa and Europe, it will benefit all partner countries and will boost their economies. However, there is a need to study climate change impacts on glaciers dynamics and lakes formation in the vicinity of CPEC to secure the route from future vulnerabilities and disasters.

Acknowledgements

This study is supported by the NSFC-ICIMOD joint project and other grants (Grant no. 41761144075 and 209071). We are thankful to the Water and Power Development Authority (WAPDA) of Pakistan for providing the required information data. We are thankful to the editor and associate editor, as well as reviewers for their valuable comments and suggestions to improve this chapter.

Conflict of interest

There is no conflict of interest.

Author details

Muhammad Saifullah[1,2], Shiyin Liu[1*], Muhammad Adnan[1,3], Muhammad Ashraf[4], Muhammad Zaman[5], Sarfraz Hashim[2] and Sher Muhammad[6]

1 Institute of International Rivers and Eco-Security, Yunnan University, Kunming, China

2 Department of Agricultural Engineering, Muhammad Nawaz Shareef University of Agriculture, Multan, Pakistan

3 State Key Laboratory of Cryospheric Science, Northwest Institute of Eco-Environment and Resources, Chinese Academy of Sciences (CAS), Lanzhou, China

4 Department of Agricultural Engineering, Khwaja Fareed University of Engineering and Information Technology, Rahim Yar Khan, Pakistan

5 Department of Irrigation and Drainage, University of Agriculture, Faisalabad, Pakistan

6 International Center for Integrated Mountain Development (ICIMOD), Kathmandu, Nepal

*Address all correspondence to: shiyin.liu@ynu.edu.cn

References

[1] Hayat H, Akbar TA, Tahir AA, Hassan QK, Dewan A, Irshad M. Simulating current and future riverflows in the snowmelt-runoff model and RCP scenarios. Water. 2019;**11**(4):761-779

[2] Immerzeel WW, Wanders N, Lutz AF, Shea JM, Bierkens MFP. Reconciling high-altitude precipitation in the upper Indus basin with glacier mass balances and runoff. Hydrology and Earth System Sciences. 2015;**19**:4673-4687

[3] Hewitt F. Woman's work, woman's place: The gendered life-world of a high mountain community in northern Pakistan. Mountain Research and Development. 1989;**9**(4):335-352

[4] Immerzeel WW, Wanders N, Lutz AF, Shea JM, Bierkens MFP. Reconciling high-altitude precipitation in the upper Indus basin with glacier mass balances and runoff. Hydrology and Earth System Sciences. 2015;**19**(11):4673-4687

[5] Tahir AA, Chevallier P, Arnaud Y, Ashraf M, Bhatti MT. Snow cover trend and hydrological characteristics of the Astore River basin (Western Himalayas) and its comparison to the Hunza basin (Karakoram region). Science of the Total Environment. 2015;**505**:748-761

[6] Bhambri R, Watson CS, Hewitt K, Haritashya UK, Kargel JS, Pratap Shahi A, et al. The hazardous 2017-2019 surge and river damming by Shispare glacier, Karakoram. Scientific Reports. 2020;**10**(1):1-14

[7] Copland L, Sylvestre T, Bishop MP, Shroder JF, Seong YB, Owen LA, et al. Expanded and recently increased glacier surging in the Karakoram. Arctic, Antarctic, and Alpine Research. 2011;**43**(4):503-516

[8] Rankl M, Kienholz C, Braun M. Glacier changes in the Karakoram region mapped by multimission satellite imagery. The Cryosphere. 2014;**8**(3):977-989

[9] Biemans H, Siderius C, Lutz AF, Ahmad B, Hassan T, Von Bloh W, et al. Importance of snow and glacier meltwater for agriculture on the indogangetic plain. Nature Sustainability. 2019;**2**(7):594-601

[10] Hewitt K. Glacier change, concentration, and elevation effects in the Karakoram Himalaya, upper Indus Basin. Mountain Research and Development. 2011;**31**(3):188-200

[11] Milner AM, Khamis K, Battin TJ, Brittain JE, Barrand NE, Olafsson S. Glacier shrinkage driving global changes in downstream systems. Proceedings of the National Academy of Sciences. 2017;**114**(37):9770-9778

[12] Westoby MJ, Glasser NF, Brasington J, Hambrey MJ, Quincey DJ, Reynolds JM. Modeling outburst floods from moraine-dammed glacial lakes. Earth-Science Reviews. 2014;**134**:137-159

[13] Lake G, Floods O. Identification of glacial flood hazards in Karakoram range using remote sensing technique and risk analysis. Science Vision. 2011;**16**:71-80

[14] Bishop MP, Kargel JS, Kieffer HH, Mackinnon DJ, Raup BH, Shroder JF. Remote-sensing science and technology for studying glacier processes in high Asia. Annals of Glaciology. 2000;**31**:164-170

[15] Shafique M, Faiz B, Sher A, Saleem B. Evaluating glacier dynamics using temporal remote sensing images: A case study of Hunza Valley, northern Pakistan. Environmental Earth Sciences. 2018;**77**(5):162

[16] Wang W, Xiang Y, Gao Y, Lu A, Yao T. Rapid expansion of glacial lakes caused by climate and glacier retreat in the Central Himalayas. Hydrological Processes. 2015;**29**(6):859-874

[17] Gilany N, Iqbal J. Simulation of glacial avalanche hazards in Shyok Basin of upper Indus. Scientific Reports. 2019;**9**(1):1-14

[18] Schwanghart W, Worni R, Huggel C, Stoffel M, Korup O. Uncertainty in the Himalayan energy-water nexus: Estimating regional exposure to glacial lake outburst floods. Environmental Research Letters. 2016;**11**(7):074005

[19] Ross L, Pérez-Santos I, Parady B, Castro L, Valle-Levinson A, Schneider W. Glacial lake outburst flood (GLOF) events and water response in a patagonian fjord. Water. 2020;**12**(1):248

[20] Bolch T, Pieczonka T, Mukherjee K, Shea J. Brief communication: Glaciers in the Hunza catchment (Karakoram) have been nearly in balance since the 1970s. Cryosphere. 2017;**11**(1):531-539

[21] Westoby MJ, Brasington J, Glasser NF, Hambrey MJ, Reynolds JM, Hassan MAAM, et al. Numerical modelling of glacial lake outburst floods using physically based dam-breach models. Earth Surface Dynamics. 2015;**3**(1):171-199

[22] Jawaid MZ. Glacial lake flood hazard assessment and modelling: A GIS perspective [MSc thesis]. 2017. p. 56

[23] Ahmad Z, Hafeez M, Ahmad I. Hydrology of mountainous areas in the upper Indus basin, northern Pakistan with the perspective of climate change. Environmental Monitoring and Assessment. 2012;**184**(9):5255-5274

[24] Ashraf A, Naz R, Roohi R. Glacial lake outburst flood hazards in Hindukush, Karakoram and Himalayan ranges of Pakistan: Implications and risk analysis. Geomatics, Natural Hazards and Risk. 2012;**3**(2):113-132

[25] Ashraf A, Roohi R, Naz R, Mustafa N. Monitoring cryosphere and associated flood hazards in high mountain ranges of Pakistan using remote sensing technique. Natural Hazards. 2014;**73**(2):933-949

[26] Cuellar AD, McKinney DC. Decision-making methodology for risk management applied to Imja Lake in Nepal. Water. 2017;**9**(8):14-16

[27] Somos-Valenzuela MA, Chisolm RE, Rivas DS, Portocarrero C, McKinney DC. Modeling a glacial Lake outburst flood process chain: The case of Lake Palcacocha and Huaraz, Peru. Hydrology and Earth System Sciences. 2016;**20**(6):2519-2543

[28] Somos-Valenzuela MA, Chisolm RE, McKinney DC, Rivas D. Inundation modeling of a potential glacial lake outburst flood in Huaraz, by High Mountains adaptation partnership inundation modeling of a potential glacial lake outburst flood in Huaraz, Peru. Center for Research in Water Resources, University of Texas at Austin. CRWR Online Report. 2014;(01)

[29] ICIMOD. Inventory of Glacial Lakes in the Koshi. China: Gandaki and Karnali River basins of Nepal and Tibet; 2018

[30] Shafique M, Faiz B, Bacha A. Evaluating glacier dynamics using temporal remote sensing images: A case study of hunza valley, northern Pakistan. The International Archives of the Photogrammetry, Remote Sensing and Spatial Information Sciences (ISPRS Archives). 2019;**42**(2/W13):1781-1785

[31] Campbell DJG, Li X, Tongliang G, Tej Partap BR, Arora BS. Inventory of Glaciers, Glacial Lakes and the Identification of Potential Glacial Lake Outburst Floods (GLOFs) Affected by Global Warming in the Mountains

of India, Pakistan and China/ Tibet Autonomous Region. Nepal: International Centre for Integrated Mountain Development; 2004

[32] King O, Bhattacharya A, Bhambri R, Bolch T. Glacial lakes exacerbate Himalayan glacier mass loss. Scientific Reports. 2019;**9**(1):1-9

[33] Gao H, Zou X, Wu J, Zhang Y, Deng X, Hussain S, et al. Post-20th century near-steady state of Batura glacier: Observational evidence of Karakoram anomaly. Scientific Reports. 2020;**10**(1):1-10

[34] Quincey DJ, Luckman A. Brief communication: On the magnitude and frequency of Khurdopin glacier surge events. The Cryosphere. 2014;**8**(2):571-574

[35] Steiner JF, Kraaijenbrink PDA, Jiduc SG, Immerzeel WW. Brief communication: The Khurdopin glacier surge revisited-extreme flow velocities and formation of a dammed lake in 2017. The Cryosphere. 2018;**12**(1):95-101

[36] Herreid S, Pellicciotti F, Ayala A, Chesnokova A, Kienholz C, Shea J, et al. Satellite observations show no net change in the percentage of supraglacial debris-covered area in northern Pakistan from 1977 to 2014. Journal of Glaciology. 2015;**61**(227):524-536

Chapter 7

Close-Range Sensing of Alpine Glaciers

Daniele Giordan, Niccolò Dematteis, Fabrizio Troilo, Valerio Segor and Danilo Godone

Abstract

Glacial processes can have a strong impact on human activities in terms of hazards and freshwater supply. Therefore, scientific observation is fundamental to understand their current state and possible evolution. To achieve this aim, various monitoring systems have been developed in the last decades to monitor different geophysical and geochemical properties. In this manuscript, we describe examples of close-range monitoring sensors to measure the glacier dynamics: (i) terrestrial interferometric radar, (ii) monoscopic time-lapse camera, (iii) total station, (iv) laser scanner, (v) ground-penetrating radar and (vi) structure form motion. We present the monitoring applications in the Planpincieux and Grandes Jorasses glaciers, which are located in the touristic area of the Italian side of the Mont Blanc massif. In recent years, the Planpincieux-Grandes Jorasses complex has become an open-air research laboratory of glacial monitoring techniques. Many close-range surveys have been conducted in this environment and a permanent network of monitoring systems that measures glacier surface deformation is presently active.

Keywords: Mont Blanc, monitoring, remote sensing, data integration, glacial hazards

1. Introduction

Mountain glaciers represent the main source of fresh water for human activities of the surrounding regions [1, 2]. Furthermore, glaciological processes (e.g. ice break-offs, glacier outbursts, snow/ice avalanches) can threaten population, urban areas and infrastructures [3]. In densely populated areas, such as the European Alps, the interaction between glaciers and anthropic activities is very frequent and it is of crucial importance to study the glaciers to understand their evolution and response to climate change, which is expected to reduce their area coverage and increase their instability [4].

Long-term monitoring of glaciological processes is often complicated and expensive, especially in remote areas and inaccessible terrains, which are common in mountain environment [5]. A practical approach is the adoption of remote sensing apparatuses that allow observing glacial processes with minimal risk for scientists and technicians. In recent years, the free availability of data acquired from satellite platforms has largely improved the possibility to observe wide areas from remote with relatively high spatiotemporal resolution. Nevertheless, satellite surveys suffer complex geometries and the revisit time might be not adequate to measure fast processes. Therefore, the use of close-range remote sensing systems is often the most effective solution for glacier monitoring [6].

Section 2 presents a substantial list of close-range remote sensing techniques that can be adopted to measure glacier surface deformations. Section 3 is devoted to the Planpincieux and Grandes Jorasses glaciers (Mont Blanc massif) case study (**Figure 1**). In recent years, such a glacial complex has become an open-air laboratory where innovative and experimental monitoring systems have been developed [6–12]. Several practical examples of close-range remote sensing surveys will be described therein.

Figure 1.
*Overview of the Planpincieux and Grandes Jorasses glaciers (upper tile) and area of study (lower tile). Yellow and orange rectangles indicate respectively the Montitaz Lobe and the Whymper Serac framed by the time-lapse cameras (**Figure 2**).*

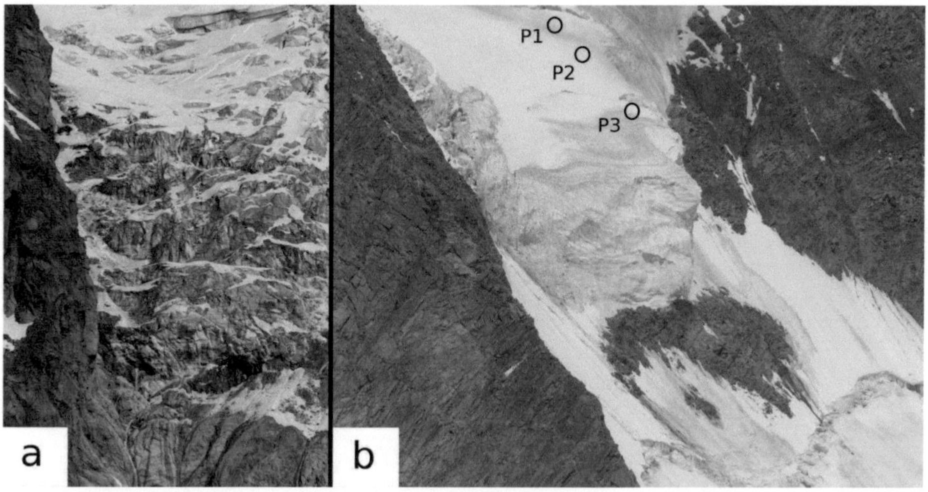

Figure 2.
(a) Image of the Montitaz Lobe of the Planpincieux Glacier monitored by a monoscopic time-lapse camera. The terminus width is approximately 100 m. (b) Image of the Whyper Serac acquired by monoscopic time-lapse camera. The serac face is approximately 40-m high. The black circles indicate the prism positions onto the serac surface in 2019.

2. Close-range remote sensing techniques

The study of the Planpincieux and Grandes Jorasses glacier surface deformations has been conducted following different approaches: (i) volumetric changes have been evaluated with point clouds and digital elevation models (DEMs). Such measurements have been obtained with laser scanners or structure from motion (SFM) processing. (ii) Surface kinematics maps of specific displacement components, which have been provided by monoscopic time-lapse camera and terrestrial interferometric radars. (iii) 3D displacements measured in specific points with a robotised total station (RTS). Furthermore, helicopter-borne ground-penetrating radar (GPR) campaigns have been conducted to investigate the glacier internal structure and thickness (**Table 1**).

2.1 Point clouds for surface generation

Three-dimensional point clouds are crucial tools in glacier monitoring; the main survey techniques to obtain them are LiDAR [13], terrestrial laser scanner (TLS) [14] and aerial and terrestrial photogrammetry, particularly structure from motion (SfM) approach [15]. LiDAR and TLS are based on a sensor, terrestrial or airborne, capable of emitting laser pulses at high frequency and measure their 'time of flight' in order to compute the position of each echo. The absolute position of each point is calculated from the emitter centre, geocoded by a GNSS coupled with an inertial measurement unit [16]. Besides its coordinates, each point can be characterised by the intensity of the echo in order to detect the nature of the target [17]. By the exploitation of laser beam divergence, it is also possible to discriminate and analyse multiple echoes or even the full waveform, thus obtaining multiple measurements of different object hit by the same pulse [18].

Concerning SfM, it is a technique originating from computer vision, which, by processing multiple images from different points of view of the same target object, generates a three-dimensional point cloud. The algorithm matches common

Glacier	Survey	Dates	References
Planpincieux	GPR	2/4/2013,	
		2019	
	DIC	August 2013-in	[8]
		course	
	TRI	9/8/2013-	[9, 42]
		10/8/2013	
		7/8/2014-	
		8/8/2014	
		1/9/2015-	
		14/10/2015	
		13/6/2016-	
		19/6/2019	
		26/9/2019-in	
		course	
	LiDAR	9/6/2014	
	TLS	2/10/2015	
	Helicopter-	2017	
	borne SFM	28/10/2018	
		20/9/2019	
		1/10/2019	
		5/11/2019	
	Drone SFM	24/7/2019	
Grandes	RTS	2010-in course	[11, 12]
Jorasses			
	DIC	2016-in course	
	GPR	4/6/2010	
		2/4/2013	
	Helicopter-	July 2010	
	borne SFM		
	Drone SFM	July 2019	

Table 1.
List of the surveys conducted in the Planpincieux and Grandes Jorasses glaciers since 2010.

features in the images and reconstructs the three-dimensional coordinates of the matching points and of the cameras. Resulting points are then collected in the cloud [19]. Images can be captured by various kinds of sensors including cameras, smartphones and drones [20].

2.2 Punctual topographic displacement measurements

Robotised total station (RTS) is a topographic apparatus that measures the sensor-to-target range and the azimuth and zenith angles, which allow determining the target position in a 3D coordinate system whose centre corresponds to the

RTS itself. Typical measurement sensibility of best-quality RTS is of 1.5 mm and 0.5 arcsec, depending on the distance. The RTS is composed of a laser rangefinder and an electronic theodolite that measures respectively distance and angles. The RTS targets retroreflector prisms installed both in and outside the moving area. The latter ones serve as control points for measurement calibration and data corrections.

Since it is required to install prisms within the investigated area, the RTS cannot be considered a remote sensing device in a strict sense. Nevertheless, such installation is needed just once; thereafter, the RTS provides measurements from remote, strongly reducing human and financial costs for accessing the surveyed area. This holds especially when the RTS works in automatic target recognition (ATR) mode, with which it carries out autonomously the measurements. In geosciences, the RTS is widely used for gravitational slope phenomena, such as landslides [21, 22], volcanos [23] and glaciers [11, 12, 24].

2.3 Glacier surface kinematics maps

Spatially distributed data are a relevant tool in glaciological studies because they allow to analyse the surface kinematic patterns and to identify possible different kinematic sectors. In the Planpincieux-Grandes Jorasses glacial complex, two main remote sensing systems have been applied to measure surface kinematics maps: digital image correlation (DIC) and terrestrial radar interferometry (TRI).

2.3.1 Digital image correlation

With the advent of digital cameras, time-lapse imagery has become popular since the beginning of the 2000s in glaciology, where it has been applied to survey polar ice flow [25–28] and mountain glaciers [6, 8, 29–33].

DIC is an image analysis technique that is applied to a pair of images to obtain spatially distributed maps of the two displacement components orthogonal to the line-of-sight (LOS). In classical DIC processing, a reference template out of the master image is searched for in an investigated larger area of the slave image. The cross-correlation (CC) is calculated for every possible template of the investigated area and the position of the maximum correlation coefficient corresponds to the displacement of the master template. Alternatively, the CC can be calculated in the Fourier domain according to the convolution theorem. Fourier CC is computationally efficient but it is more prone to outliers.

The main DIC advantages concern the low-cost hardware and its high portability even in harsh environments. Nevertheless, it suffers adverse meteorology and it strongly depends on the visibility conditions.

2.3.2 Terrestrial radar interferometry

In the last two decades, TRI revealed to be a valuable tool to monitor glaciers [9, 34–42]. TRI concerns the analysis of the phase difference between two radar acquisitions, which is directly related to the target displacement component parallel to the LOS. Typical radar apparatuses can provide spatially distributed displacement data in an area of several square kilometres with an operative range of a few kilometres. Radars are active sensors, as such, TRI can be applied during the night and severe meteorological conditions. Moreover, TRI measurements have sub-millimetre sensibility in an optimal context. However, the processing is not trivial and it requires high computational costs. Particularly complicated is the phase wrapping solution, which depends on the phase 2 periodicity and which is related to the sensor-to-target range. Moreover, TRI is quite sensitive to possible morphological change of the scattering surface and that causes signal decorrelation and extreme atmospheric conditions can heavily affect the measurements [43, 44]. In glaciological contexts, long distances, morphological surface changes and severe meteorology are common and TRI processing must be handled carefully.

2.4 Glacier internal structure

GPR has been widely used as a geophysical method for the study of internal glacier properties. A variation in electrical permittivity creates dielectric interfaces and subsequent reflections that can be analysed. GPR can be used for the definition of firn-ice transition, the detection of subglacial cavities and the ice thickness [45]. GPR systems include a transmitter and a receiver antenna. Typical operating frequencies vary between 10 and 15 MHz, for the investigation of glaciers having depths of hundreds of meters, to 400–600 MHz, for shallow investigations. Different factors can limit the effectiveness of the technique, such as debris cover of the ice surface or highly crevassed areas that can create scattering or absorption phenomena that reduce the possibility of investigation of the glacier sub-surface. Processing of radar data normally implies many steps, which include (i) low-frequency filtering, caused mainly by surface reflection; (ii) selection of a time gain to correct for the amplitude divergence; (iii) temporal and spatial filtering for improving the signal-to-noise ratio; (iV) deconvolution and (v) migration [46].

GPR apparatuses are usually lightweight and compact and they can be easily transported by walking or snowmobile, which allows at acquiring a large number of 2D radar profiles. However, helicopter-borne surveys provide the most versatile platform and they have been used for detecting glacier thickness [47, 48], intraglacial features [49] and snow accumulation [50].

2.5 Data integration

Spatially distributed deformation data provide wide information on the investigated process. Nevertheless, common remote sensing apparatuses only provide specific displacement components or punctual measurements and the integration of different sensors is necessary to obtain spatially distributed 3D data.

Dematteis et al. [6] proposed an innovative solution to obtain 3D displacement using DIC and TRI data integration. DIC and TRI provide different and complementary displacement components that can be coupled to obtain a three-dimensional representation of the surface kinematics. The necessary conditions to couple the different data are that their maps must have the same spatial resolution in the same coordinate system (CS). Therefore, a geometric transformation is required to represent both data in the same CS, which is usually associated with a georeferenced DEM.

A different approach of data integration entails the merging of DIC and RTS data. RTS provides 3D displacement in specific points, while DIC can measure spatially distributed data. Therefore, their integration allows obtaining the displacement direction and versus using RTS data, while the DIC results give the spatial distribution.

3. Case study: Planpincieux-Grandes Jorasses glaciers

The Planpincieux and Grandes Jorasses glaciers form a unique polythermal glacial complex located on the Italian side of the Grandes Jorasses peak (Mont

Blanc massif), in the Ferret valley (**Figure 1**). The glaciers have approximately a South-East aspect and the elevation ranges from 2600 m asl to 4200 m asl. The accumulation area of the Grandes Jorasses Glacier is formed of two 45° steep cirques, which merge in an icefall at 3500 m asl. In the left cirque is located the Whymper Serac, whose front is at an elevation of 3800 m asl (**Figure 2b**). According to Pralong and Funk [51], this portion is classified as an unbalanced hanging glacier. As such, the serac progressively increases its volume and when its shape reaches unstable geometry, the serac collapses. This cycle follows an irregular periodicity and the time between the break-offs ranges from a few years to more than a decade. Usually, the unstable ice chunk has a volume of the order of 105 m^3, which can collapse at once or in several pieces. The instability dynamics is driven only by the geometry and it is not linked to temperature or water percolation. Therefore, the fracture can also occur during the cold season, when the collapse might easily trigger a large snow avalanche that would seriously threaten the underlying buildings and the road at the valley bottom. The last events happened in August 1993, June 1998 [11] and September 2014 [12]. The first one caused the fatality of eight mountaineers, but the ice avalanche did not cause further damages for the absence of snow.

The Planpincieux Glacier topography presents three distinguished zones: the accumulation area, 3000–3500 m asl, is formed of two steep cirques that merge in a wide plateau at 2900–3000 m asl, and two lobes constitute the ablation area. The right lower lobe (**Figure 2a**) is 32° steep on average and it is quite crevassed. Its terminus ends in correspondence of a bedrock cliff that causes frequent calving. In the past, several collapses occurred and, in a few cases, they endangered the bridge of the Montitaz stream that originates from the glacier snout. Further information on the Planpincieux Glacier can be found in Giordan et al. [7].

3.1 Monitoring campaigns

In the last decades, the Planpincieux-Grandes Jorasses Glacier complex has become an open-air laboratory where innovative remote sensing techniques have been developed to monitor the glacier activity [6–12].

The Planpincieux Glacier is observed by two monoscopic time-lapse cameras placed in the opposite side of the Ferret valley, at a distance of 3800 m from the glacier. The monitoring station is equipped with two solar panels and an electric cell for power supply. It is remotely controlled by a Raspberry Pi 3 connected to the server of the Geohazard Monitoring Group (GMG) of the Research Institute for Geo-Hydrological Protection (IRPI), in Torino, Italy. A robotised webcam has been installed in 2018 to survey the station functioning. The system is active since August 2013 and it acquires images at hourly frequency. In the period August 2013–December 2019, it collected more than 35,000 images and it is probably the longest continuous series of hourly images in the European Alps. The images are processed with the DIC technique to estimate the surface glacier kinematics.

The Grandes Jorasses Glacier is being monitored since 2010 by an RTS installed in the Planpincieux hamlet at a distance of 4800 m. The RTS measures every 2 h the position of the prisms installed onto and in the vicinity of the Whymper Serac (**Figure 2b**). Due to the extreme meteorological conditions and the exceptional sensor-to-target range, the prisms are not always visible and gaps in the measurement series are frequent, especially during the cold season. Snowfalls and strong wind occasionally cause the loss of some prisms and the intervention of Alpine guides it is necessary for the installation of new targets. Moreover, the Whymper Serac is continuously monitored by a 4800-m-far monoscopic camera. This survey is active since 2010 and the serac surface displacement is estimated with feature tracking of the hourly photographs.

Besides these continuous monitoring systems, in the past, several measurement campaigns have been conducted to increase the glacier understanding and to develop new monitoring techniques of glaciological close-range remote sensing. In **Table 1**, the complete list of the surveys conducted since 2010 is presented and the related references are reported when available.

3.1.1 Point cloud analysis

DEMs obtained during LiDAR and TLS surveys and from photographic SfM acquired by drones and helicopter-borne cameras allow monitoring the morphology evolution of the glacier surface. In addition, the DEM of difference (DoD) calculation permits to estimate the surface elevation changes and the possible ice mass loss. From the DoD obtained with the DEMs acquired in October 2019 and June 2014 (helicopter-borne SfM and LiDAR respectively), one can observe the glacier thinning of more than 10 m on average (**Figure 3**). In the considered period, the terminus retreated by several tenths of metres and the bedrock remained exposed. In this part, the DoD shows a thickness loss of 30–40 m approximately, which corresponds to the glacier thickness in 2014.

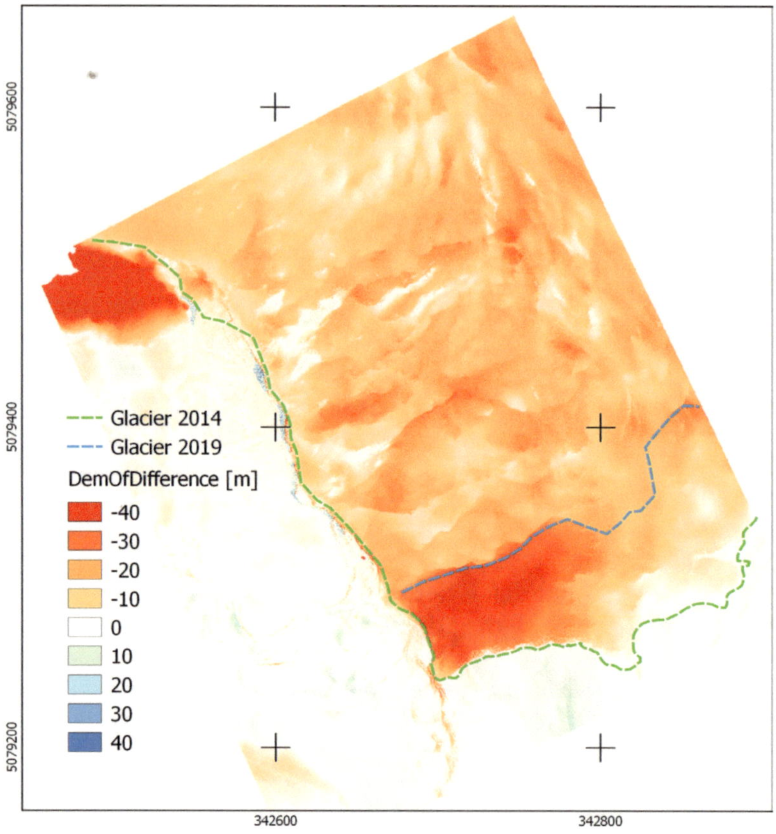

Figure 3.
DEM of difference (DoD) of the Montitaz Lobe. The DoD is calculated as the difference between the DEMs acquired on 1/10/2019 and 9/6/2014. The glacier outlines in both years are represented as dashed lines.

3.1.2 RTS applications

RTS measurements are continuously active since 2010 to monitor the surface velocity of the Whymper Serac. The survey is conducted with a Leica TM30 that operates in ATR mode. The prism network is composed of several stakes installed into the unstable portions, while a few prisms placed in the surrounding bedrock serve as reference points. Complete acquisition of the entire network lasts approximately 45 min and it is conducted every 2h. The sensor-to-target distance is of 4800 m on average, which is beyond the instrument operating limits declared by the manufacture in ATR mode (https://w3.leica-geosystems.com/downloads123/zz/tps/tm30/brochures-datasheet/tm30_technical_data_en.pdf). In addition, extreme atmospheric conditions linked to the high-mountain elevation occur frequently. This situation makes the Whymper Serac a critical scenario for RTS measurements and a robust processing method has been developed ad hoc [11]. However, the RTS data allowed forecasting 10 days in advance the serac break-off of 22/10/2014 [12]. The RTS data acquired before such an event are shown in **Figure 4**.

3.1.3 Time-lapse camera applications

The surface kinematics of the Planpincieux Glacier right lobe has been deeply investigated with image analysis of 6-year-long time-lapse monitoring. The data analysis allowed characterising the terminus dynamics and classifying the instability processes that cause break-offs: (i) disaggregation, (ii) slab fracture and (iii) water tunnelling [7]. Disaggregation is the progressive toppling of small ice pieces caused by the movement of the terminus beyond the frontal bedrock cliff. It is the most frequent process and it involves break-offs of limited size, usually lower than 1000 m^3. Slab fracture instability is caused by the aperture of a crevasse orthogonal to the motion direction, located in correspondence to the maximum tensile stress line. When the fracture reaches the bedrock, it triggers a large break-off of an ice lamella that can assume a volume of 104–105 m^3. Water tunnelling refers to the formation of R-channels [52] where a large amount of

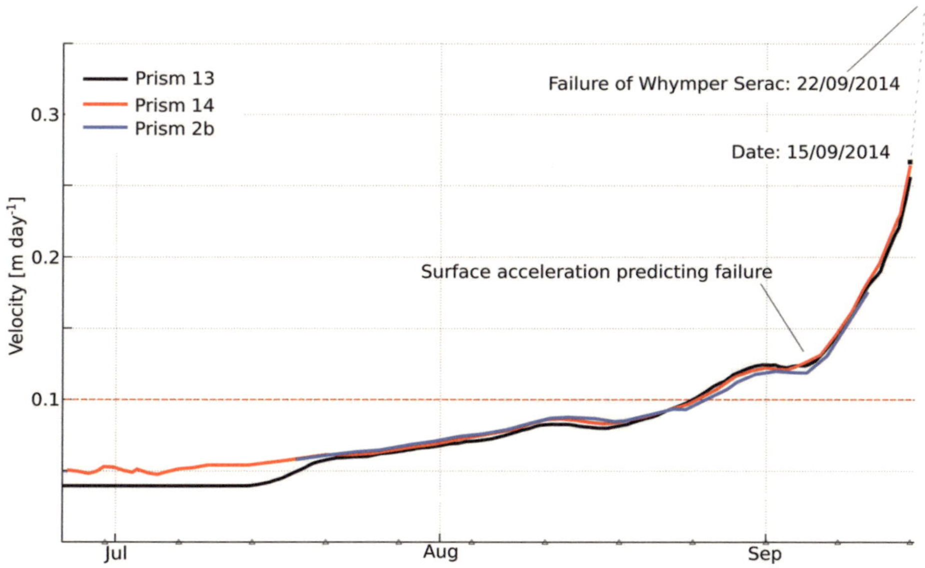

Figure 4.
RTS measurements of prisms 13, 14, and 2b before the failure of 22/09/2014. Using these data, the break-off was predicted 10 days in advance.

water can accumulate. The water produces a strong pressure on the frontal cliff that can provoke failure of the terminus. Moreover, the empty tunnels increase the instability and they can collapse themselves.

Besides the visual photographic interpretation, DIC in the Fourier domain was applied to the hourly images, obtaining surface displacement maps at daily resolution. During the monitoring period, the surface displacement pattern was composed of four distinct kinematic domains, which were characterised by different velocity regimes. The presence of kinematics domains indicates the action of high strain rates localised at the domain limits, where large fractures appear (**Figure 5**). The behaviour of the frontal sector is noteworthy, because it reveals the occurrence of a few speed-up periods per year, which culminate with large break-offs (**Figure 6**). These kinematic fluctuations were characterised by well-defined thresholds of initial velocity ($v_0 \geq 30$ cm day^{-1}) and acceleration ($a \geq 3$ cm day^{-2}). Moreover, a monotonic relationship (rank correlation coefficient > 0.7, p-value < 0.02) between the velocity peak and the collapsed volume has been observed.

DIC in the spatial domain was applied to the images of the Whymper Serac to measure the displacement in July 2019 (**Figure 7a**, **b**). The available images presented rototranslation that had to be compensated with robust coregistration. Moreover, the smooth texture and low chromatic contrast of the scene lowered the signal-to-noise ratio (i.e. the correlation, see **Figure 7**) and hence many artefacts were present in the displacement maps. Therefore, a robust outlier correction method was applied [53]. The results showed a slight acceleration during July, which was confirmed by RTS measurements.

3.1.4 TRI applications

The Planpincieux is probably the unique glacier where TRI surveys were conducted using four different terrestrial interferometric radar models, namely: GPRI™ (Gamma Remote Sensing, https://www.gamma-rs.ch/rud/microwave-hardware/gpri.html), IBIS-L™ (IDS Georadar, https://idsgeoradar.com/products/interferometric-radar/ibis-fl), FastGBSAR-S™ (MetaSensing, https://www.geomatics.metasensing.com/fastgbsar-s) and GBInSAR LiSALab™ (LiSALab, http://www.lisalab.com/home/default.asp?sez=6).

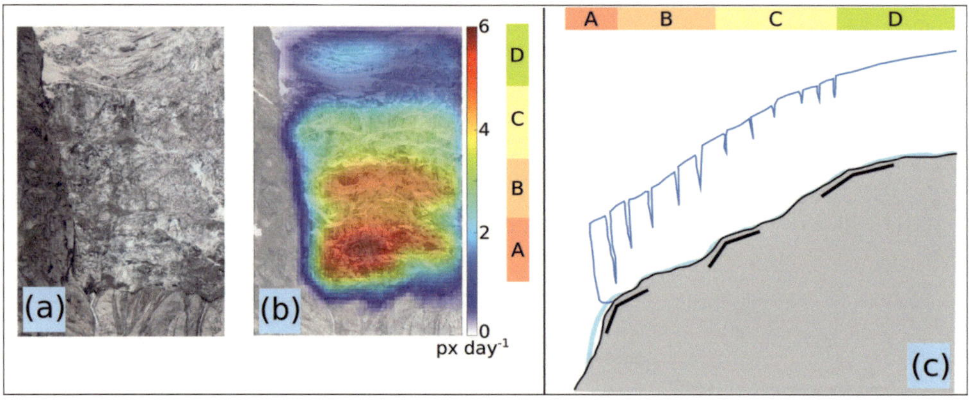

Figure 5.
(a) Image of the Montitaz Lobe acquired by the monoscopic time-lapse camera. The terminus width is approximately 100 m. (b) Surface deformation map. Different velocity regimes clearly identify the four kinematics domains. (c) Longitudinal conceptual scheme of the glacier lobe (not in scale). The black lines indicate bedrock discontinuities that correspond to the kinematic domain limits. Modified from Giordan et al. [7].

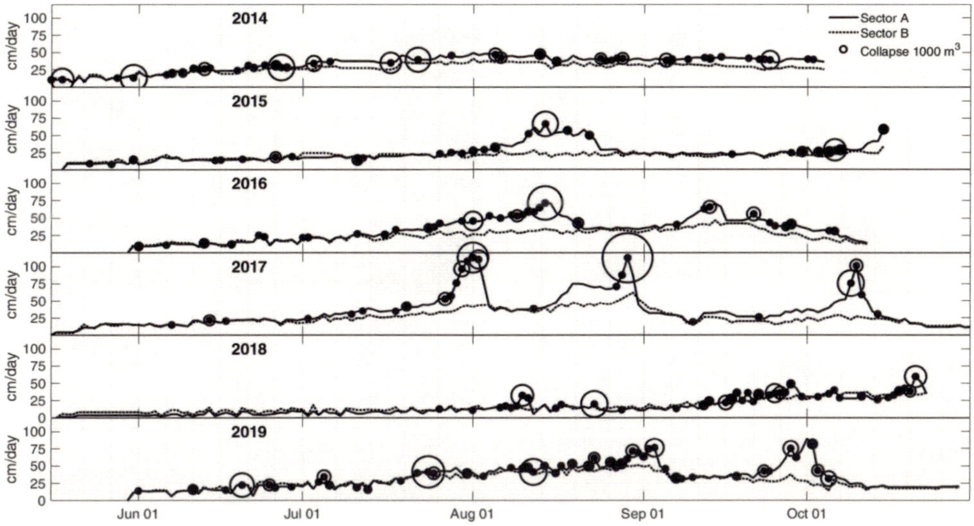

Figure 6.
*Time series of the daily velocity of sectors A and B (see **Figure 5**) in the years 2014–2019 (from top to bottom). The break-off occurrence is depicted in black dots, while the white circle size is proportional to the volume.*

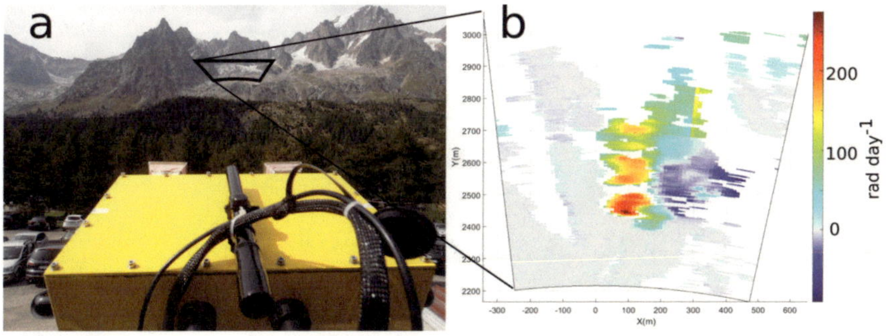

Figure 7.
(a) IBIS-L GBSAR surveyed the Planpincieux Glacier area, delimited in black. (b) The cumulative sum of the interferograms acquired in the period September 4, 2015 to October 14, 2015.

The surface kinematics of the glaciers was surveyed in five TRI campaigns, in 2013, 2014, 2015, 2016 and 2019 (**Table 1**). The first two were conducted using the GPRI™ real-aperture radar (RAR) in Ku band that surveyed the glacier from the valley bottom and the valley ridge opposite to the glaciers. Both campaigns lasted for 2 days and they were able to detect the displacements of the lower portions of the Planpincieux and Grandes Jorasses glaciers, which were approximately 25 cm day^{-1} and 50 cm day^{-1} respectively. Instead, the following surveys were conducted using Ku-band ground-based synthetic aperture radars (GB-SAR). The campaign of autumn 2015 (IBIS-L™) lasted much longer and hence it was possible to recognise the different kinematic domains of the Montitaz Lobe (**Figure 8**). During the campaign, the meteorological conditions were severe and the radar acquisitions were affected by strong APS. To solve the issue, a polynomial APS model that was a function of the topography was developed [9, 42]. In 2016, FastGBSAR-S™ measurements with an acquisition frequency of 10 s were carried out; thereby, the atmospheric disturbance was minimised. Fully polarimetric measurements were experimented, but the very long distance did not allow exploiting the potentiality of such a technology. The last campaign (GBInSAR LiSALab™) began at the end of September 2019 for civil protection and it is still active during the writing of the present chapter.

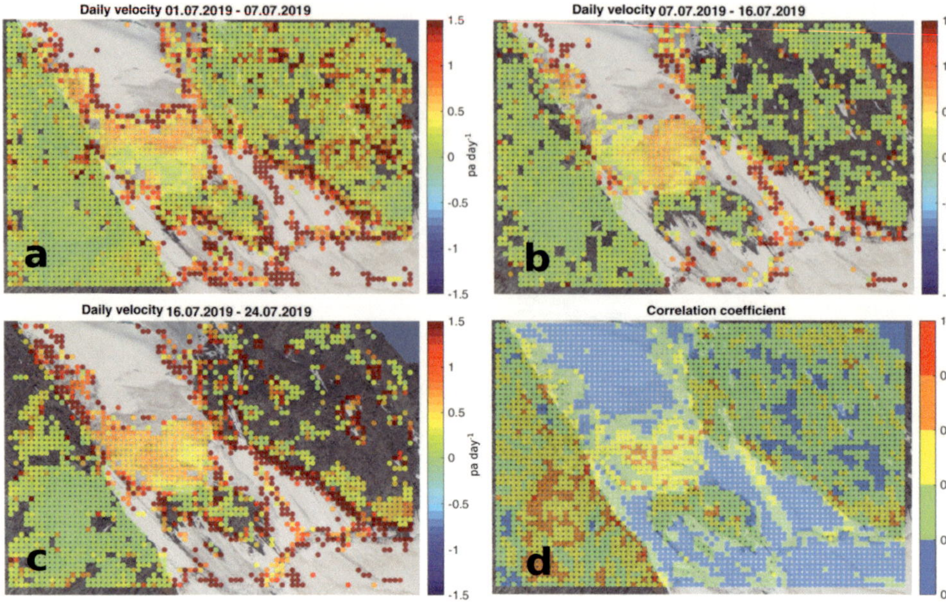

Figure 8.
(a-c) Surface displacement maps of the Whymper Serac of the periods July 1, 2019 to July 7, 2019, July 7, 2019 to July 16, 2019 and July 16, 2019–July 24, 2019. (d) Map of the mean correlation coefficient, which displays low values because of the texture smoothness of the snow surfaces. The serac face is approximately 40 m high.

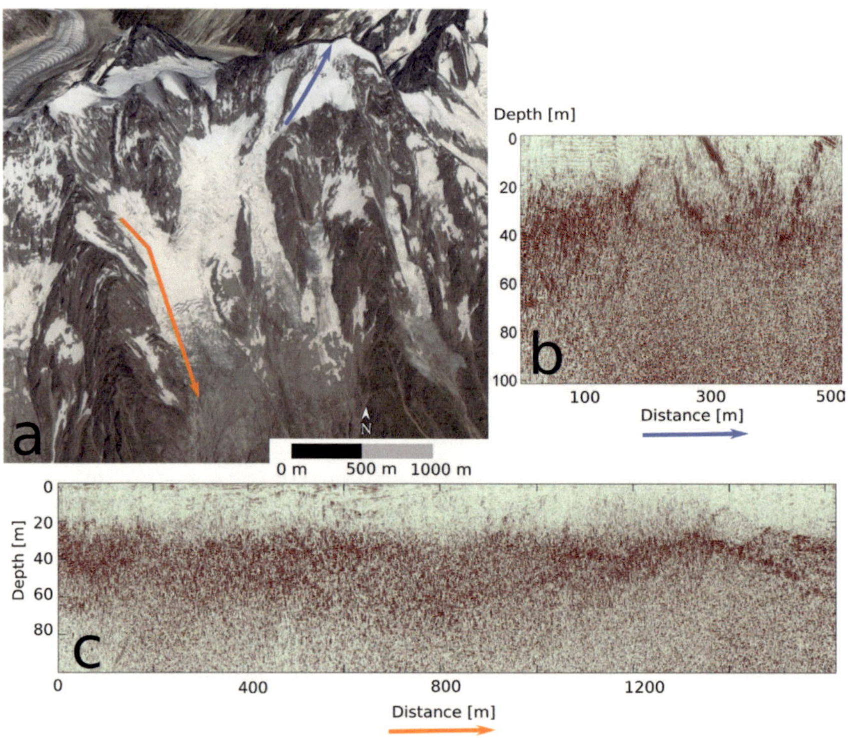

Figure 9.
(a) GPR traces of the Planpincieux (orange line) and Grandes Jorasses (blue line) glaciers. (b-c) GPR profiles of the Whymper Serac and Planpincieux Glacier respectively. The white-red boundary indicates the ice thickness.

3.1.5 GPR applications

A helicopter-borne 65-MHz GPR survey was conducted in the Planpincieux-Grandes Jorasses glacial complex in April 2014, when 16 GPR traces homogeneously distributed on the glaciers' surface were acquired (**Figure 9**). The noise of the radar data was quite high, because the numerous crevasses caused bounds of the electromagnetic waves and produced echoes and artefacts. Nevertheless, it was possible to estimate the glacier thickness, which was in the range 20–40 m in the Planpincieux Glacier and lower than 20 m in the Whymper Serac.

3.1.6 Data integration

In September 2015, time-lapse photography and terrestrial radar campaigns were conducted simultaneously to measure the Planpincieux Glacier surface kinematics. The actual three-dimensional surface kinematics was obtained by coupling DIC and TRI results. **Figure 10** reports the mean daily velocity map, where the colour represents the velocity module and the arrows indicate direction and versus. The 3D displacement can be obtained only in the areas visible by both the sensors. In the right lobe, the displacement vectors are not uniformly parallel to the surface, because the seracs move downstream as a single body and the ice is subjected to internal deformation. This result is not trivial, as the most common approach to estimate 3D displacement is to project the single movement components along the local slope obtained from the DEM, but this assumption might be misleading in specific cases.

The permanent monitoring system of the Whymper Serac is composed of RTS and time-lapse imagery. In July 2019, the data of the two sensors were integrated and represented in an informative bulletin [54, 55], shown in **Figure 11**. Such integration allows evaluating the versus and direction of the principal movement (with the RTS data) and the distribution of the strain rates (with the DIC results).

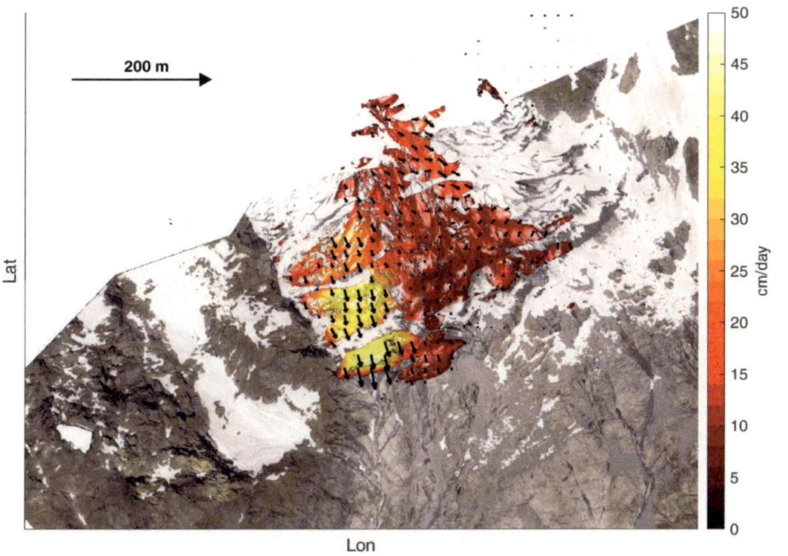

Figure 10.
Velocity field of the surface kinematics of the lower Planpincieux Glacier obtained with the integration of DIC and TRI measurements. Colours and arrows represent velocity module and direction respectively. Modified from Dematteis et al. [6].

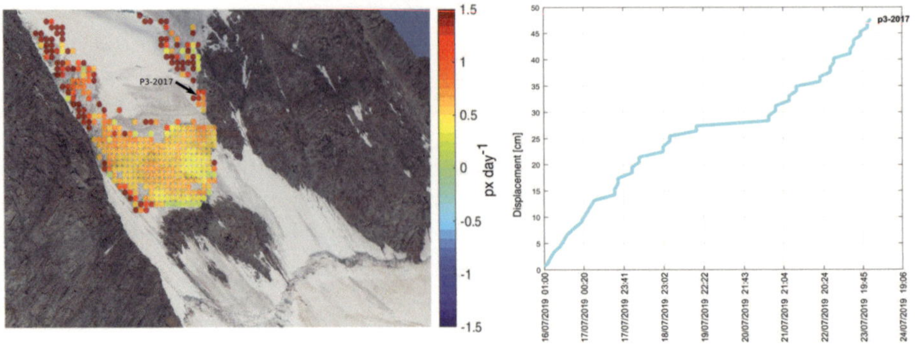

Figure 11.
Data integration of DIC and RTS measurements. The image depicts the spatially distributed daily deformation of the Whymper Serac front (coloured dots) and the surface displacement direction measured by the TRS in correspondence with the prism P3-2017. The right plot reports the displacement trend provided by the RTS.

4. Summary

In-depth knowledge of glacier behaviour is fundamental for glaciological risk evaluation and management and it permits to develop mitigation and adaptation strategies against the cryosphere change provoked by global warming. To achieve this aim, data collection about the current glacier state is of primary importance, but the harsh mountain environment makes the survey activities difficult. Measurements from aerospace platforms are affected by complex geometries and might not provide sufficient spatiotemporal resolution, especially when high acquisition rates (i.e. minutes to hours) are necessary. Therefore, ground-based systems are often the most suitable solution. Nevertheless, impervious areas where glaciers are usually located entail the use of high financial and human efforts, as well as potential risks to access the investigated area. Therefore, remote sensing systems represent the best cost-benefit ratio and they are commonly adopted for glacier monitoring. Considering the possible adverse conditions (e.g. extreme meteorology, steep slopes, long sensor-to-target distance, natural hazards) that can occur during the survey activities, ad hoc technologies and methods must be developed. The glacial complex formed of the Planpincieux and Grandes Jorasses glaciers represents an outstanding site where different close-range remote sensing approaches have been experimented, in a heterogeneous Alpine glacier environment. Here, the combined use of multiple sensors proved to be a valuable tool to collect complementary information that allowed improving the understanding of the current state and recent evolution of the glacial area.

Author details

Daniele Giordan[1], Niccolò Dematteis[1*], Fabrizio Troilo[2], Valerio Segor[3] and Danilo Godone[1]

1 Research Institute for Geo-Hydrological Protection, National Research Council of Italy, Torino, Italy

2 Safe Mountain Foundation, Courmayeur, Italy

3 Struttura Assetto Idrogeologico Bacini Montani, Regione Autonoma Valle d'Aosta, Aosta, Italy

*Address all correspondence to: niccolo.dematteis@irpi.cnr.it

References

[1] Barnett TP, Adam JC, Lettenmaier DP. Potential impacts of a warming climate on water availability in snow-dominated regions. Nature. 2005;**438**:303-309. DOI: 10.1038/nature04141

[2] Hock R, Jansson P, Braun LN. Modelling the Response of Mountain Glacier Discharge to Climate Warming. In Global Change and Mountain Regions (A State of Knowledge Overview). Dordrecht: Springer; 2005. pp. 243-252. DOI: 10.1007/1-4020-3508-x_25

[3] Kääb A, Huggel C, Fischer L, Guex S, Paul F, Roer I, et al. Remote sensing of glacier- and permafrost-related hazards in high mountains: An overview. Natural Hazards and Earth System Sciences. 2005;**5**(4):527-554. DOI: 10.5194/nhess-5-527-2005

[4] Deline P, Gruber S, Delaloye R, Fischer L, Geertsema M, Giardino M, et al. Ice loss and slope stability in high-mountain regions. In: Snow and Ice-Related Hazards, Risks, and Disasters. Academic Press; 2014. pp. 521-561. DOI: 10.1016/B978-0-12-394849-6.00015-9. ISBN: 9780123964731

[5] Kenner R, Phillips M, Limpach P, Beutel J, Hiller M. Monitoring mass movements using georeferenced time-lapse photography: Ritigraben rock glacier, western Swiss Alps. Cold Regions Science and Technology. 2018;**145**:127-134. DOI: 10.1016/j.coldregions.2017.10.018

[6] Dematteis N, Giordan D, Zucca F, Luzi G, Allasia P. 4D surface kinematics monitoring through terrestrial radar interferometry and image cross-correlation coupling. ISPRS Journal of Photogrammetry and Remote Sensing. 2018;**142**:38-50. DOI: 10.1016/j.isprsjprs.2018.05.017

[7] Giordan D, Dematteis N, Allasia P, Motta E. Classification and kinematics of the Planpincieux Glacier break-offs using photographic time-lapse analysis. Journal of Glaciology. 2020;**66**(256):188-202. DOI: 10.1017/jog.2019.99

[8] Giordan D, Allasia P, Dematteis N, Dell'Anese F, Vagliasindi M, Motta E. A low-cost optical remote sensing application for glacier deformation monitoring in an alpine environment. Sensors. 2016;**16**(10):1750. DOI: 10.3390/s16101750

[9] Dematteis N, Luzi G, Giordan D, Zucca F, Allasia P. Monitoring Alpine glacier surface deformations with GB-SAR. Remote Sensing Letters. 2017;**8**(10):947-956. DOI: 10.1080/2150704X.2017.1335905

[10] Dematteis N, Giordan D, Allasia P. Image classification for automated image cross-correlation applications in the geosciences. Applied Sciences. 2019;**9**(11):2357. DOI: 10.3390/app9112357

[11] Margreth S, Faillettaz J, Funk M, Vagliasindi M, Diotri F, Broccolato M. Safety concept for hazards caused by ice avalanches from the Whymper hanging glacier in the Mont Blanc Massif. Cold Regions Science and Technology. 2011;**69**(2-3):194-201. DOI: 10.1016/j.coldregions.2011.03.006

[12] Faillettaz J, Funk M, Vagliasindi M. Time forecast of a break-off event from a hanging glacier. The Cryosphere. 2016;**10**(3):1191-1200. DOI: 10.5194/tc-10-1191-2016

[13] Wehr A, Lohr U. Airborne laser scanning—An introduction and overview. ISPRS Journal of Photogrammetry and Remote Sensing. 1999;**54**(2-3):68-82. DOI: 10.1016/S0924-2716(99)00011-8

[14] Godone D, Godone F. The Support of Geomatics in Glacier Monitoring: The Contribution of Terrestrial Laser Scanner. Rijeka: IntechOpen; 2012. DOI: 10.5772/33463

[15] Westoby MJ, Brasington J, Glasser NF, Hambrey MJ, Reynolds JM. "Structure-from-Motion" photogrammetry: A low-cost, effective tool for geoscience applications. Geomorphology. 2012;**179**:300-314. DOI: 10.1016/j.geomorph.2012.08.021

[16] Baltsavias EP. Airborne laser scanning: Basic relations and formulas. ISPRS Journal of Photogrammetry and Remote Sensing. 1999;**54**(2-3):199-214. DOI: 10.1016/S0924-2716(99)00015-5

[17] Glennie CL, Carter WE, Shrestha RL, Dietrich WE. Geodetic imaging with airborne LiDAR: The Earth's surface revealed. Reports on Progress in Physics. 2013;(8):24-76. DOI: 10.1088/0034-4885/76/8/086801

[18] Chauve A, Mallet C, Bretar F, Durrieu S, Pierrot-Deseilligny M, Puech W, et al. Processing full-waveform Lidar data: Modelling raw signals. In: Proceedings of the International Archives of Photogrammetry Remote Sensing and Spatial Information Sciences. 2007. pp. 102-107

[19] Fonstad MA, Dietrich JT, Courville BC, Jensen JL, Carbonneau PE. Topographic structure from motion: A new development in photogrammetric measurement. Earth Surface Processes and Landforms. 2013;**38**:421-430. DOI: 10.1002/esp.3366

[20] Cignetti M, Godone D, Wrzesniak A, Giordan D. Structure from motion multisource application for landslide characterization and monitoring: The champlas du col case study, sestriere, north-western Italy. Sensors (Switzerland). 2019;**19**(10):2364. DOI: 10.3390/s19102364

[21] Manconi A, Allasia P, Giordan D, Baldo M, Lollino G, Corazza A, et al. Landslide 3D surface deformation model obtained via RTS measurements. In: Landslide Science and Practice. Berlin, Heidelberg: Springer Berlin Heidelberg; 2013. pp. 431-436. DOI: 10.1007/978-3-642-31445-2_56

[22] Allasia P, Baldo M, Giordan D, Godone D, Wrzesniak A, Lollino G. Near real time monitoring systems and periodic surveys using a multi sensors UAV: The case of Ponzano landslide. In: IAEG/AEG Annual Meeting Proceedings, San Francisco, California, 2018-Volume 1. Springer International Publishing: Cham; 2018. pp. 303-310. DOI: 10.1007/978-3-319-93124-1_37

[23] Langbein JO. Deformation of the Long Valley Caldera, California: Inferences from measurements from 1988 to 2001. Journal of Volcanology and Geothermal Research. 2003;**127**(3-4):247-267. DOI: 10.1016/S0377-0273(03)00172-0

[24] Nainwal HC, Negi BDS, Chaudhary M, Sajwan KS, Gaurav A. Temporal changes in rate of recession: Evidences from Satopanth and Bhagirath Kharak glaciers, Uttarakhand, using Total Station Survey. Current Science. 2008;**94**(5):653-660

[25] Ahn Y, Box JE. Glacier velocities from time-lapse photos: Technique development and first results from the Extreme Ice Survey (EIS) in Greenland. Journal of Glaciology. 2010;**56**(198):723-734. DOI: 10.3189/002214310793146313

[26] Dietrich R, Maas HG, Baessler M, Rülke A, Richter A, Schwalbe E, et al. Jakobshavn Isbræ, West Greenland: Flow velocities and tidal interaction of the front area from 2004 field observations. Journal of Geophysical Research: Earth Surface. 2007;**112**(3):F03S21. DOI: 10.1029/2006JF000601

[27] Brinkerhoff D, O'Neel S. Velocity variations at Columbia Glacier captured by particle filtering of oblique time-lapse images. arxiv.org; 2017

[28] Schwalbe E, Maas HG. The determination of high-resolution spatio-temporal glacier motion fields from time-lapse sequences. Earth Surface Dynamics. 2017;**5**(4):861-879. DOI: 10.5194/esurf-5-861-2017

[29] Benoit L, Dehecq A, Pham HT, Vernier F, Trouvé E, Moreau L, et al. Multi-method monitoring of Glacier d'Argentière dynamics. Annals of Glaciology. 2015;**56**(70):118-128. DOI: 10.3189/2015AoG70A985

[30] Messerli A, Grinsted A. Image georectification and feature tracking toolbox: ImGRAFT. Geoscientific Instrumentation, Methods and Data Systems. 2015;**4**(1):23-34. DOI: 10.5194/gi-4-23-2015

[31] Fallourd R, Trouvé E, Roşu D, Vernier F, Bolon P, Harant O, et al. Monitoring temperate glacier displacement by multi-temporal TerraSAR-X images and continuous GPS measurements. IEEE Journal of Selected Topics in Applied Earth Observations and Remote Sensing. 2011;**4**(2):372-386. DOI: 10.1109/JSTARS.2010.2096200

[32] Vernier F, Fallourd R, Friedt JM, Yan Y, Trouvé E, Nicolas J-M, et al. Fast correlation technique for glacier flow monitoring by digital camera and space-borne SAR images. EURASIP Journal on Image and Video Processing. 2011;**1**:11. DOI: 10.1186/1687-5281-2011-11

[33] Evans AN. Glacier surface motion computation from digital image séquences. IEEE Transactions on Geoscience and Remote Sensing. 2000;**38**(2 II):1064-1072. DOI: 10.1109/36.841985

[34] Allstadt KE, Shean DE, Campbell A, Fahnestock M, Malone SD. Observations of seasonal and diurnal glacier velocities at Mount Rainier, Washington, using terrestrial radar interferometry. The Cryosphere. 2015;**9**(6):2219-2235. DOI: 10.5194/tc-9-2219-2015

[35] Luzi G, Pieraccini M, Mecatti D, Noferini L, Macaluso G, Tamburini A, et al. Monitoring of an alpine glacier by means of ground-based SAR interferometry. IEEE Geoscience and Remote Sensing Letters. 2007;**4**(3):495-499. DOI: 10.1109/LGRS.2007.898282

[36] Noferini L, Mecatti D, Macaluso G, Pieraccini M, Atzeni C. Monitoring of Belvedere Glacier using a wide angle GB-SAR interferometer. Journal of Applied Geophysics. 2009;**68**(2):289-293. DOI: 10.1016/j.jappgeo.2009.02.004

[37] Riesen P, Strozzi T, Bauder A, Wiesmann A, Funk M. Short-term surface ice motion variations measured with a ground-based portable real aperture radar interferometer. Journal of Glaciology. 2011;**57**(201):53-60. DOI: 10.3189/002214311795306718

[38] Voytenko D, Dixon TH, Werner C, Gourmelen N, Howat IM, Tinder PC, et al. Monitoring a glacier in southeastern Iceland with the portable terrestrial radar interferometer. In: Proceedings of the International Geoscience and Remote Sensing Symposium (IGARSS). 2012. pp. 3230-3232. DOI: 10.1109/IGARSS.2012.6350736

[39] Voytenko D, Stern A, Holland DM, Dixon TH, Christianson K, Walker RT. Tidally driven ice speed variation at Helheim Glacier, Greenland, observed with terrestrial radar interferometry. Journal of Glaciology. 2015;**61**(226):301-308. DOI: 10.3189/2015JoG14J173

[40] Xie S, Dixon TH, Voytenko D, Holland DM, Holland D, Zheng T. Precursor motion to iceberg calving at Jakobshavn Isbræ, Greenland,

observed with terrestrial radar interferometry. Journal of Glaciology. 2016;**62**(236):1134-1142. DOI: 10.1017/jog.2016.104

[41] López-Moreno JI, Alonso-González E, Monserrat O, Del Río LM, Otero J, Lapazaran J, et al. Ground-based remote-sensing techniques for diagnosis of the current state and recent evolution of the Monte Perdido Glacier, Spanish Pyrenees. Journal of Glaciology. 2019;**65**(249):85-100. DOI: 10.1017/jog.2018.96

[42] Luzi G, Dematteis N, Zucca F, Monserrat O, Giordan D, López-Moreno JI. Terrestrial radar interferometry to monitor glaciers with complex atmospheric screen. In: Proceedings of the International Geoscience and Remote Sensing Symposium (IGARSS); Vol. 2018, July. 2018. pp. 6243-6246. DOI: 10.1109/IGARSS.2018.8519008

[43] Caduff R, Schlunegger F, Kos A, Wiesmann A. A review of terrestrial radar interferometry for measuring surface change in the geosciences. Earth Surface Processes and Landforms. 2015;**40**(2):208-228. DOI: 10.1002/esp.3656

[44] Monserrat O, Crosetto M, Luzi G. A review of ground-based SAR interferometry for deformation measurement. ISPRS Journal of Photogrammetry and Remote Sensing. 2014;**93**:40-48. DOI: 10.1016/j.isprsjprs.2014.04.001

[45] Pellikka P, Rees W. Remote Sensing of Glaciers: Techniques for Topographic, Spatial and Thematic Mapping of Glaciers. Boca Raton, USA: CRC Press; 2009. ISBN: 978-0-415-40166-1

[46] Daniels DJ. Ground Penetrating Radar: Theory and Applications. 2nd ed. The Institution of Electrical Engineers, London; 2004

[47] Macheret YY, Zhuravlev AB. Radio echo-sounding of Svalbard glaciers. Journal of Glaciology. 1982;**28**(99):295-314. DOI: 10.1017/S0022143000011643

[48] Damm V. Ice thickness and bedrock map of Matusevich Glacier drainage system (Oates Coast). Terra Antart. 2004;**11**(1-2):85-90

[49] Arcone SA, Yankielun NE. 1.4 GHz radar penetration and evidence of drainage structures in temperate ice: Black Rapids Glacier, Alaska, USA. Journal of Glaciology. 2000;**46**(154):477-490

[50] Machguth H, Eisen O, Paul F, Hoelzle M. Strong spatial variability of snow accumulation observed with helicopter-borne GPR on two adjacent Alpine glaciers. Geophysical Research Letters. 2006;**33**(13). DOI: 10.1029/2006GL026576

[51] Pralong A, Funk M. On the instability of avalanching glaciers. Journal of Glaciology. 2006;**52**(176):31-48. DOI: 10.3189/172756506781828980

[52] Röthlisberger H. Water pressure in intra- and subglacial channels. Journal of Glaciology. 1972;**11**(62):177-203. DOI: 10.3189/s0022143000022188

[53] Hart DP. The elimination of correlation errors in PIV processing. In: Proceedings of the 9th International Symposium on Applications of Laser Techniques to Fluid Mechanics; Volucella. 1998. pp. 13-16

[54] Giordan D, Wrzesniak A, Allasia P. The importance of a dedicated monitoring solution and communication strategy for an effective management of complex active landslides in urbanized areas. Sustainability. 2019;**11**(4):946. DOI: 10.3390/su11040946

[55] Wrzesniak A, Giordan D. Development of an algorithm for automatic elaboration, representation and dissemination of landslide monitoring data. Geomatics, Natural Hazards and Risk. 2017;**8**(2):1898-1913. DOI: 10.1080/19475705.2017.1392369

Chapter 8

Glacial Biodiversity: Lessons from Ground-dwelling and Aquatic Insects

Mauro Gobbi and Valeria Lencioni

Abstract

At first glance, the ground surrounding the glacier front and the streams originated by melting glaciers seem to be too extreme to host life forms. They are instead ecosystems, colonized by bacteria, fungi, algae, mosses, plants and animals (called the "glacial biodiversity"). The best adapted animals to colonize glacier surface, the recently deglaciated terrains and glacial streams are insects, specifically the ground beetles (carabids) and the non-biting midges (chironomids). This chapter aims to overview the species colonizing these habitats, their adaptation strategies to face natural cold and anthropogenic heat and the extinction threats of glacial retreat and pollution by emerging contaminants. Notes on their role in the glacial-ecosystem functioning and related ecosystem services are also given.

Keywords: carabid beetles, chironomids, cold-adapted species, debris-covered glaciers, extinction risk, glacier forelands, rock glaciers

1. Introduction

Insects are the most diverse and abundant group of animals on Earth and are critical drivers of ecosystem function in terrestrial and aquatic systems.

Biodiversity of insects is threatened worldwide [1]; 40% of the world's insect species could go extinct within decades [2] with consequent loss of ecosystem functions and services.

Despite the attention from the media, and scientific community, it remains unclear whether such declines are widespread among habitats and geographic regions. Since most of the insects are providing services (e.g., pollination and decomposition) and disservices (e.g., damaging crops and spreading diseases), the efforts in insect conservation biology and population management have been mainly conveyed to highly impacted areas like, for instance, the lowlands.

On the other hand, around the world, mountain regions are changing at an unprecedented rate. Most of the evidences are based on the abiotic component (e.g., temperature increase and precipitation variations), but there are increasing evidences about changes in biological communities.

At high altitude in tropical and temperate mountains and at high latitude, habitat loss, pollution and climate change affect negatively cold-adapted insect species distribution and survival [3, 4]. It is unlikely that insect declines will be homogenous everywhere, but some general patterns can be identified. For example, at high altitude, the high frequency of extreme climatic events and the loss of ice-related landforms (e.g., glaciers and rock glaciers) are detrimental to cold-adapted insects, highly specialized to survive constantly at low temperature, high humidity, deep snow and ice cover. Most of the glaciers are strongly reducing their surface, some are disappearing, others are shifting in debris-covered glaciers and permafrost is melting [5]. The effects of climate change on high-altitude terrestrial and aquatic insect community composition and ecosystem functioning are partially unknown and still underconsidered [6].

Scattered studies on high-altitude and high-latitude insect communities are available for most of the glacialized areas of the world, but a synthesis paper merging the most recent advances on cold-adapted insect species living on and at the edge of ice-related landforms is still not available. To fill this gap, the purpose of this chapter is to (i) describe what insects live in glacialized areas, (ii) examine what are the flagship insects in terms of richness and adaptation to cold, (iii) identify threats and opportunities in the current warm-period and (iv) illustrate the role of insect communities in the glacial-ecosystem functioning and services.

2. Glacial biodiversity: glacial and periglacial landforms as habitat for cryophilous species

At first glance, the ground surrounding the glacier front, its surface and the streams originated by the melting glaciers seems to be too extreme to host any life forms. Conversely, there are many organisms that permanently colonize the cryosphere, from bacteria to vertebrates, the so-called glacial biodiversity [7].

From the quantitative point of view, the glacial biodiversity is mainly formed by little (less of 5 cm) or microscopic organisms linked to aquatic, semiaquatic, wet or terrestrial (mainly bare and rocky grounds) habitats.

The surface of glaciers is colonized by bacteria and algae [8, 9] and mosses, hosting in turn water bears (Tardigrade), roundworms (Nematoda) and potworms (Enchytraeidae) [10]. They can live also directly on the surface of the ice, in cryoconite holes and rivulets on ice. In addition, on the glaciers, it is possible to observe wandering several spiders (Araneae) during the day as well as some ground beetles (Coleoptera: Carabidae) during the night searching for preys. Many other invertebrates, mostly connected with aquatic/wet environments, such as springtails (Collembola), stoneflies (Plecoptera) and non-biting midges (Diptera: Chironomidae), can live directly on the surface of the ice [11]. Moreover, supraglacial cryoconite holes can be considered biodiversity hotspots for invertebrates such as Copepoda, worms (Annelida), water bears, roundworms and wheel animals (Rotifera) [11, 12].

Most of the invertebrates living on the glaciers can be found also along recently deglaciated terrains and in the uppermost reach of glacial streams. Thus, the glacial biodiversity living at the edge with the glacier front arrives from two different sinks: falling down from the glacier surface during the melting phases and shifting up along the glacier forelands, transported airborne or by walking. Typically, springtails and moss mites (Oribatida) are the first colonizers of recently deglaciated areas because they are easily transported by the wind, thanks to their lightweight. Some sheet weaver spiders (Linyphiidae) are as well colonizers transported by wind [13]. Thus, recently deglaciated terrains act as a source of glacial biodiversity characterized by species living on the ground but still linked to the presence of cold microclimate ensured by the presence of the glacier few meters far from them.

Aquatic fauna in rivulets on ice (= eukryal) and uppermost reaches of glacier-fed streams (=metakryal) is represented mainly by cold hardy non-biting midges that have resistant forms to survive freezing and desiccation. Other guests of such habitats are worms (e.g., Naididae and Enchytraeidae) and free-living and parasite roundworms.

Similar to recently deglaciated terrains, also rock glaciers—which are the best expression of the periglacial environment in alpine areas—can sustain glacial biodiversity, also in the places where there are no glaciers or they disappeared. Currently, the state of knowledge about the animals living on this kind of landform is still very limited; just few sites in Sierra Nevada and on the Alps were investigated [14–16]. Both studies highlighted the presence of microbes, arthropods and small rodent fauna linked to the cold and wet habitat ensured by the presence of interstitial ice. Given the known thermal and hydrologic capacity of rock glaciers to resist warming, this distinct landform has been suggested as potential climatic refugia for glacial biodiversity in the current warm stage period.

It is therefore evident that glacial biodiversity is characterized by species living in ice-related landforms, thus landforms ensuring, at the ground as well as on the water, average annual temperatures around 0°C [17, 18]. Thus, the species living on and around ice-related landforms can be defined as cryophilous (which means "loving the ice") or cold-adapted species. Among these, the best adapted to survive such extreme habitats are ground beetles (carabids) and non-biting midges (chironomids).

3. Ground beetles and non-biting midges as flagship organisms: richness and adaptation

Ground beetles (**Figure 1**) occur in almost all terrestrial environments and geographical areas and have different trophic requirements (e.g., predators, seed eaters and phytophagous), it is easy to collect them and their sensitivity to environmental and climate changes is known [19]. In addition, they can be considered among the most important meso- and macrofauna living on recently deglaciated terrains and on the glaciers in terms of species richness and abundance of individuals [13]. Last but not least, there is an ongoing awareness about the extinction risk for some endemic cryophilous species due to the habitat destruction (e.g., glacier

Figure 1.
The ground beetle Nebria germari *(approx. body length = 1 cm) walking on the Presena Glacier (2700 m a.s.l.) (Adamello-Presanella Mts., Italian Alps) (photo by F. Pupin/archive MUSE).*

disappearing) or changes in microhabitat conditions (e.g., permafrost melt). Most of the species living at high altitudes and latitudes have low dispersal abilities due to the lack of wings and are walking colonizers, ground hunters and small-sized, which are traits typical of species living in cold, wet and gravelly habitats [17]. To date, about 40,000 species are known in the world [20] and most of them are endemic to specific areas [21–23]; for instance, about 28% of the total species belonging to the Italian carabid fauna are endemic.

Non-biting midges (**Figure 2**) are the freshwater insect family that comprises the highest number of species, both in lentic and lotic habitats [24]. They are the most widespread of all aquatic insect families, with individual species occurring from Antarctica to the equator lands and the Arctic, from lowlands to thousands of meters of altitudes. There are species that thrive in almost every conceivable freshwater environment. Ice-cold glacial trickles, hot springs, thin films, minute containers of water in the leaf axils of plants and the depths of great lakes all have their characteristic species or communities. There are semiaquatic species, living in moist soil or vegetation and others that are truly terrestrial with few species occurring in marine water. Some species tolerate brackish water, others thrive in intertidal pools and, unusually among the insects, and a few are truly marine. Survival in harsh environments is due to a series of adaptations. Among these are the production of melanin, their small size, capacity for mating on the ground instead of in flight (they therefore have small or totally absent wings), the building of cocoons, diapause and resistance to cold [25]. To date, about 6500 species are known in the world; one tenth of which are in Italy and one thousandth in Alpine streams, springs and lakes.

Thanks to their species richness, adaptation to cold environments, key role in the ecological network structure and robustness and sensitivity to short-term and long-term climate changes, ground beetles and non-biting midges might be considered flagship organisms of the glacial biodiversity.

3.1 Ground beetles on glaciers

Clean glaciers and debris-covered glaciers can host permanent populations of ground beetles, at least on the European glaciers since, to our knowledge, there are no data from other extra-European mountain chains. All the species found on the European glaciers belong to the genera *Nebria* and *Oreonebria*. The *Nebria/*

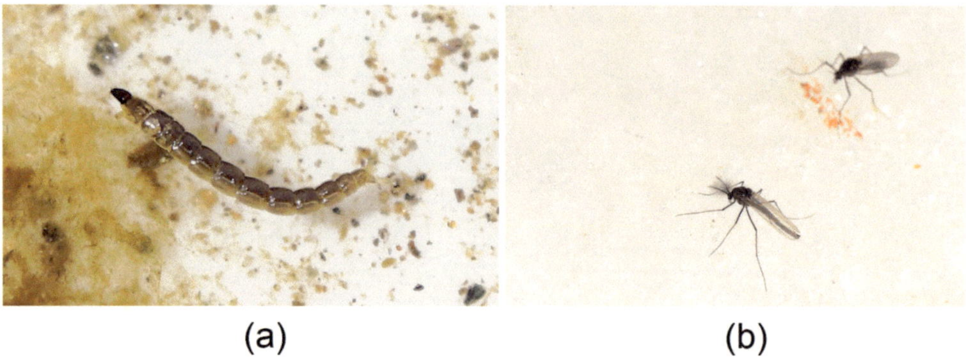

Figure 2.
The non-biting midge Diamesa zernyi *larva (a) from the Amola glacial stream (2540 m a.s.l.) and a couple of adults (male on the left, female on the right) (b) walking on the Presena Glacier (2700 m a.s.l.) (Adamello-Presanella Mts., Italian Alps) (photo by V. Lencioni) (a) and F. Pupin/archive MUSE (b). Animal sizes ca 1 cm.*

Oreonebria species living on clean glaciers are with reduced and not functional wings and wander on the glacier mainly during the night searching for preys (mainly springtails, spiders, non-biting midges and died insects). During the day, they find refuge under the rocks on glacier surface and within the moraines. Their legs are longer with respect to those closely-related species living at lower altitudes and in different habitats, in order to maintain the body to a higher distance from the frozen ground [26].

In addition, debris-covered glaciers are able to host permanent population of *Nebria/Oreonebria* species (**Figure 3**). Currently, data are available for five debris-covered glaciers of the Italian Alps [17, 18, 27, 28].

Nebria/Oreonebria are olfactory-tactile predators; it means that they use the chemoreceptors located on the antenna as instrument to find preys on the glacier surface or between the stony debris. Therefore, these organisms are well adapted to move between the stony debris covering the glaciers, thus across a three-dimensional space. The sex ratio on the glacier is female-biased [29], and the colonization of the glacier from the neighboring habitat seems be done by females that have a higher propensity to disperse than males [29].

3.2 Ground beetles along glacier forelands

The gradual melting of glaciers leave in front of them large areas of barren, pristine ground open for colonization of various life forms. Among these, ground beetles can be found along the entire glacier foreland. Ground beetles can colonize entire glacier forelands, from sites deglaciated since more than one-hundred years to sites deglaciated since one year (**Figure 4**).

The colonization of a glacier foreland by ground beetles is triggered mainly by time since glaciation, distance to glacier and vegetation cover, as highlighted by studies carried out in Northern Europe (e.g., [30–32]), Alps (e.g., [27, 33–36]) and more recently Andes [37]. The colonization of a glacier foreland by ground beetles can follow two different models: the "addition and persistence" and "replacement-change" models [31]. The former was mainly observed in Northern-Europe and on Andes [31, 37], with an exception in the peripheral mountain range of the Southern Alps [38]. It consists in the persistence of pioneer species (i.e., the initial colonizers, e.g., *Nebria* spp. in Europe; *Dyscolus* spp. on the Andes [37]) from the sites deglaciated few years ago (early successional stages) to sites deglaciated more than 100 years ago (late successional stages)—in this case, there is no species turnover along the chronosequence of glacier retreat. The "replacement-change" model, mainly observed on the Alps (e.g., [17, 36]), consists in a group of initial colonizers (the pioneer community) progressively replaced over time by one or more other species assemblages; thus, in this case, there is a clear species turnover. Notwithstanding these different models of colonization, Northern Europe and Alps share ground beetles belonging to common genera and exhibiting the same patterns of colonization. For instance, the species belonging to the genus *Nebria* are surface active predators able to immediately colonize deglaciated terrains of the European glacier forelands. The species belonging to the genus *Amara* and *Carabus*, the former omnivorous and the latter specialized predators, arrived on a deglaciated terrain after more than 20 years [30, 33]. A quite common pattern observed along the European glacier forelands is that the number of species increases with the time since deglaciation, with a more diversified community on terrains deglaciated 100 years ago.

The speed of colonization along the glacier forelands varies with the time since deglaciation. Specifically it is high in the first years after the glacier retire due to the low competition in colonizing pristine terrains, while it is low in terrains located far from the glacier front because more competitive in terms of microhabitat and resource availability [35]. In the contest of the ongoing climate warming, it is interesting to highlight that an increase of 0.6°C in summer temperatures approximately doubled the speed of initial colonization, whereas later successional stages were less sensitive to climate change [39].

Figure 3.
Stony debris covering the surface of the Sorapiss Centrale glacier (45°N, 12°E, Ampezzo Dolomites, Italian Alps); it hosts permanent populations of cold-adapted ground beetles, spiders and springtails (photo by M. Gobbi).

Figure 4.
Pitfall trap near the front of the Agola Glacier (46°N, 10°E, Brenta Dolomites, Italy). Pitfall trapping is one of the most successful methods to collect ground-dwelling invertebrates at high altitudes during the snow free period (photo by M. Gobbi).

3.3 Ground beetles on rock glaciers

Currently, data on ground beetles found on rock glaciers are available only for the Italian Alps [16, 40 41]. Active rock glaciers are a unique landform: the occurrence of permafrost and the size of the stones differentiate them from the surrounding landforms (e.g., scree slopes) in terms of temperature regime and depth of the substrate. Active rock glaciers show occurrences of cold-adapted species. Even though ground beetle communities of active rock glaciers show few differences in terms of species richness and abundance with respect to scree slopes, some characteristic species of each of the two landforms can be identified. The ground beetle community observed on the rock glaciers is exclusive of this landform because it is composed of large populations of species belonging to the genera *Oreonebria*, *Nebria* and *Trechus* [16, 40]. To these genera belong species (e.g., *Nebria germari*, *Oreonebria soror* and *Trechus tristiculus*) typical of cold and wet high-altitude environments. These species have two kinds of life style: epigeic (they move on the surface of the rock glacier where the rocky detritus is fine) and endogeic (they reach the depth of the stony detritus moving between the interstitial space between stones). Conversely, the surrounding ice-free landforms (e.g., scree slopes) host species assemblages characterized by the presence of species typical of alpine grasslands (e.g., *Carabus* spp. and *Cymindis vaporariorum*). Therefore, an active rock glacier can be defined as a superficial subterranean habitat [16] represented by fissure network among boulders, human-sized caves included.

Unlike other superficial subterranean habitats like scree slopes, where temperatures could reach relatively high values in summer [16, 42], rock glaciers are selected by cold-adapted species, which avoid scree slopes as they do not offer constantly low temperatures during summer.

3.4 Non-biting midges on the glacier

To our knowledge, the only aquatic insects found permanently colonizing the ice are non-biting midges of the genus *Diamesa* in temperate zones and the stonefly *Andiperla willinki* (family Gripopterygiidae) in South America [43]. Larvae of *Diamesa steinboecki* and *Diamesa latitarsis* were collected on one Alpine glacier (2625–2650 m a.s.l., Agola, Brenta Dolomites, Italy), surviving a summer temperature ranging for 0.07 to 0.19°C. Larvae of *Diamesa* were collected also on Yala glacier (5100–5700 m a.s.l., Nepal, Himalayas [44, 45]), growing in melt-water drainage channels under the ice and feeding on blue-green algae and bacteria. They eat the scares allochthonous detritus transported by the wind and left by the glacier in the ice melt waters. Typically, primary food resources in eukryal consist of dust (allochthonous particles or airborne detritus) and algae (various species of cyanophytes and green algae), and fungi and bacteria associated with algae and detritus [46]. Adults were brachypterous (characterized by reduced wings), unable to fly, walking at temperatures as low as –16°C on the surface of the glacier and in small cavities beneath it.

3.5 Non-biting midges in proglacial ponds

Non-biting midges are first colonizers of ephemeral ponds in the proximity of glacier snout. The appearance of new ponds is usually followed by their rapid disappearance and by a concomitant appearance of new ones, frequently observed in the Alps above the tree line [47]. Most of them are relatively small (surface <2 ha) [48], unproductive due to their sparse soil development and small catchments. Ponds are especially susceptible to the effects of climatic changes because of their relatively low water volumes and high surface area-to-depth ratios. Therefore, they act as early indicators of the impacts of climate change. During the ice-free months, typically, they undergo high-level fluctuations due to ice-snow melt rate and rainfall pattern. In winter, they freeze, totally if shallow [48]. Water temperature can be highly variable as well, ranging from 0 to 15°C during summer.

The zoobenthic community of Alpine proglacial ponds is dominated by chironomid Orthocladiinae (generally representing >70% of the community) followed by Diamesinae (*Pseudokiefferiella parva* and *Diamesa* spp.). Aquatic beetles (e.g., Elmidae, Dytiscidae and Hydrophilidae), Oligochaeta (e.g., Enchytraeidae) and Hydracarina frequently represent the remaining fauna. Overall, the richness is low, with few dozens of species colonizing the same pond. Few taxa might be found with high density, up to >1000 individuals/m^2 in ponds >2 ha and less than 2 m deep. Among orthoclads, semiterrestrial genera are frequent (e.g., *Metriocnemus*, *Smittia* and *Parasmittia*), being environments that undergo high water level fluctuations during the ice-free period [49] (**Figure 5**). There are evidence of colonization by up to 4–5 congeneric species of *Metriocnemus* in

Figure 5.
Catching Metriocnemus *adults (non-biting midges) with tweezers on the shoreline of the Agola proglacial pond (2596 m a.s.l., 46°N, 10°E, Brenta Dolomites Mts., Italian Alps) (photo by D. Debiasi/archive MUSE).*

single Alpine ponds (Agola glacier, 2596 m a.s.l., Brenta Dolomites, Italy), with *M. fuscipes* and *M. eurynotus* as dominant species. The genus is considered semiterrestrial, found in mosses, phytotelmata, springs, ditches, streams and occasionally in the middle of lakes and rock pools [50]. Some ability to survive desiccation and hibernation often in combination with cocoon building and migration of larvae into the sediment [51] has been recorded for several *Metriocnemus* species dwelling in ephemeral habitats that seasonally dry or freeze out. The colonization of these ponds by *Metriocnemus* might be due more to these physiological adaptations than to repeated recolonization as observed for other chironomids colonizing ephemeral ponds [52]. In fact, due to the high geographical isolation of the pond and scarce connectivity with other suitable habitats in the catchment, we can suppose that these species persist by activating a physiological response to physical stress. Most of these species are univoltine, entering diapause in a desiccated-frozen state until spring thawing.

3.6 Non-biting midges in glacier-fed streams

Non-biting midges are the main colonizers of glacier-fed streams around the world. Glacially dominated rivers are characterized by a deterministic nature of benthic communities due to the overriding conditions of low water temperature, low channel stability, low food availability and strong daily discharge fluctuations associated to glacier runoff (**Figure 6**). A predictable longitudinal pattern of taxa richness and diversity increasing with distance from the glacier has been described for many European glacier-fed streams, starting from the kryal sector (where maximum water temperature is below 4°C), typically colonized almost exclusively by *Diamesa* species in the temperate regions [53]. *D. steinboecki*, *D. goetghebueri*, *D. tonsa*, *D. zernyi* and *D. bohemani* are the species more frequent and abundant in kryal sites in the Palearctic regions, followed by *D. bertrami*, *D. latitarsis*, *D. modesta*, *D. hamaticornis* and *D. cinerella*. Less frequent are *D. martae*, *D. nowickiana*, *D. longipes*, *D. wuelkeri* and *D. aberrata*; *D. insignipes*, *D. dampfi*, *D. permacra* and

Figure 6.
Dubani glacial stream at the glacier snout (3232 m a.s.l., 36°N, 74°E; Bagrote Valley, Karakoram range of northern Pakistan) (photo by L. Latella).

D. incallida are rare, being more frequent in glacio-rhithral and krenal habitats, characterized by a lower glacial influence. *D. arctica* is typical of the Arctic regions, and *D. akhrorovi*, *D. alibaevae*, *D. planistyla, D. solhoyi*, *D. aculeate*, *D. praecipua* and *D. khumbugelida* are among the *Diamesa* species typical of Pamir and Tibet mountains. In tropical streams, mainly Podonominae colonize the uppermost glacier-fed stream reaches, while Diamesinae appear more downstream [54]. In glacier-fed streams of New Zealand, [55] reported the mayflies (Ephemeroptera) *Deleatidium cornutum* and *Nesameletus* dominated at the upper sites of glacier-fed streams, with the chironomids *Eukiefferiella* and *Maoridiamesa*. Biodiversity naturally increases with decreasing altitude and increasing distance from the glacier terminus, with a more diversified community downstream of the confluence with tributaries fed by groundwater and rainfall (= krenal and kreno-rhithral). Orthocladiinae and Tanypodinae (e.g., *Zavrelimyia*) become more abundant, followed by Tanytarsini (e.g., *Micropsectra* spp.) in slow-flowing waters where mosses are abundant.

4. Threats and opportunities for ground beetles and non-biting midges in relation to climate warming

Glaciers and permafrost are disappearing all over the world, and with them, we are risking to lose also the associated glacial biodiversity. Therefore, it is mandatory to describe the temporal and spatial biological fingerprint of climate change impacts to deeply understand trends and patterns.

The available literature on ground beetles and non-biting midges is able to give us insights about the threats and opportunities they have in relation to the ongoing climate and, consequently, landscape changes.

4.1 Extinction

Currently, no ground beetles living on glacial and periglacial landforms have been declared extinct. On the other hand, the temporal reduction in population size of two high-altitude species (*Nebria germari* and *Trechus dolomitanus*) of the Dolomites (Italy) in 30 years was documented [56]. Specifically, local extinction of *Nebria germari* populations was documented in some high-altitude prairies of the Dolomites, and now the species maintain large populations only on glacial and periglacial landforms; thus, it has become an ice-related species.

As observed for ground beetles, also for non-biting midges from kryal habitats, there is no evidence of global extinction of single species, rather of local extinction caused by the retreat of glaciers. The consequence of the glacial retreat is the further isolation of the populations in the short-term and, in the long-term, their possible disappearance due to very restricted habitat preference and limited dispersal abilities of midges. Glacier shrinking favors an upstream shift of lowland euriecious species of chironomid and other invertebrates, associated with an initial decrease in abundance and finally local extinction of kryal *Diamesa* species and other Diamesinae [57]. *Diamesa longipes* and *Syndiamesa nigra* have not been collected in recent years in Alpine running water [53], and the ice fly *Diamesa steinboecki* has disappeared in some glacier-fed streams in the Southern Alps [58]. The strong cold hardiness of *Diamesa* species [59] and the scarcity of potential refuge areas in glacial and periglacial area threaten these species seriously with extinction. Thus, *Diamesa* species have been suggested to be used as sentinels for climate change, especially in relation to glacier retreat. Recent studies found a direct relationship between the loss of *Diamesa* species in alpine riverine environments and the consequences of the changing climate [58].

4.2 Uphill and upstream shift

Higher temperatures and increased drought are leading to an upward shift of stenothermal species that depend on low temperatures and therefore to the fragmentation and progressive reduction in their habitat. Any endemic species, like several high-altitude ground beetles, that is restricted to summit areas and has a low dispersion ability is forced to move upward searching for microclimates suitable for its survivor. Data on ground beetles resampled in the same places after decades suggest common trend in cryophilous species. For instance, on the Andes, from 1880 to 1985, the species *Dyscolus diopsis* has shifted approximately 300 m upward, with the resulting area reduction of more than 90% from >12 km^2 to <1 km^2 [60]. The same altitudinal shift was observed on the Dolomites for the species *Nebria germari* [27] from 1950 to 2019; the habitat preference for this species was alpine prairies [61], N-exposed scree slopes and recently deglaciated terrains, and currently, it seems to be restricted only to ice-related landforms and scree slopes with high snow cover temporal extent.

Shrinking glaciers are resulting in the lengthening of glacial streams, with consequent upstream migration of specialist species to colonize the "new" stream reach, still harsh, in front of the glacier terminus. Downstream generalist species also migrate upstream, to conquer sites with ameliorated environmental conditions associated to a reduced glacial runoff and increased temperature and channel stability [62, 63]. For example, in the Alps, as first colonizers upstream were observed grazer (chironomid Orthocladiinae among which *Eukiefferiella* spp., *Heleniella* spp., *Orthocladius frigidus* and *Chaetocladius* spp.) and shredder insects (Nemouridae), covering distances from 300 m to about 2 km and a difference in altitude up to 600 m probably favored by higher amount of debris from the banks [58].

4.3 Adaptation

To the best of our knowledge, there is no evidence of physiological or morphological adaptation of carabid beetles in relation to the climate change at high altitudes, and it seems that limits to species distributions reflect present environmental tolerance limits rather than simply an historical lack of opportunity for range expansion [64]. Some studies on thermal tolerance highlighted that temperature gradients and acute thermal tolerance do not support the hypothesis that physiological constraints drive species turnover with elevation [65].

Cold stenothermal non-biting midges that adapted to live at temperatures close to their physiological limits like *Diamesa* spp. might only survive and reproduce if they can adapt to new environmental conditions or if they are able to avoid the stressor adopting specific behaviors. Barring these abilities, they are expected to disappear. There are evidences of physiological adaptation in Diamesinae to increasing water temperature in glacier-fed streams. For example, *Diamesa zernyi*, *Diamesa tonsa* and *Pseudodiamesa branickii* are cold hardy with a thermal optimum below 6°C but survive short-term heat shock by developing a heat shock response based on the synthesis of heat shock proteins [66]. It is clearly not sufficient to preserve the species considering the observed cases of local extinction. Decreasing glacier cover disadvantages Diamesinae and other cold stenothermal taxa but favors organisms with long life cycles (univoltine) or more (semivoltine) due to continuous growth around the year (life cycle shifts suggest that where glacier cover is high, nondiapausal organisms typically develop rapidly in the spring/summer melt seasons before rivers dry up or freeze through winter) [67]. Furthermore, decreasing glacier cover favors insects that undergo incomplete metamorphosis, such as Plecoptera (stoneflies) and Ephemeroptera (mayflies), and noninsect taxa such as Oligochaeta (worms), burrowing and using interstitial habitat. Dietary shifts reflect terrestrial vegetation succession with decreasing glacier cover supplying plant litter to rivers resulting in higher amount of organic material for detritivores. These shifts were observed in glacialized systems in European Alps (Austria and Italy), French Pyrénées, Greenland, Iceland, New Zealand Alps, Norway Western Fjords, US Rockies, Alaska and Svalbard [67].

4.4 Refuge areas

If the speed of adaptive capacity—when possible—is not temporally synchronous with the speed of the glacier retreat, the only way to survive for cryophilous species is to find refuge areas. A refuge can be defined as sites able to preserve suitable climate conditions for cold-adapted species in spite of the climate warming [68]. The role of active rock glaciers and debris-covered glaciers as potential warm-stage refugia for cold-adapted ground beetle species is supported by data collected on the Italian Alps [16, 28, 40]. The thermal profile observed on some alpine active rock glaciers supports this view indicating decoupling of the local topoclimate from the regional climate, a key factor for a site to serve as a refugium. Specifically, active rock glaciers differ from the surrounding landforms (e.g., scree slopes) by overall lower ground surface temperature (average annual temperatures around or below 0°C). During postglacial periods, cold-adapted species found refuge in cooler habitats, such as subterranean environments (e.g., caves), where they could find cold and stable microclimatic conditions [69]. Thus, we cannot exclude that the same pattern is acting till now for ground beetles on active rock glaciers and debris-covered glaciers, because only these landforms are still able to support large-size populations of cold-adapted species.

In streams, the majority of invertebrates avoid the hazards of freezing or desiccation (due to freezing of the substrate or due to drought caused by increasing temperature) by migrating to unfrozen habitats (e.g., springs fed by groundwater inputs and hyporheic zone), where they remain active [70]. This is a temporary adaptation, to escape daily or seasonal risk of freezing or desiccation. On long time scale, these refugia cannot preserve cold stenothermal *Diamesa steinboecki* and similar species, never found in springs and not confined to the hyporheic having the terrestrial adult. Rock glacier outflows might act as a cold refuge areas after the glacier loss also for aquatic insects due to their constantly cold waters [71]. Ref. [72] investigated five streams fed by rock glaciers in South Tyrol (Italy) and found a dominance of Diamesinae and Orthocladiinae chironomids, besides Plecoptera, Ephemeroptera and Trichoptera (EPT). The authors reported the presence of cold-stenothermal species (*Diamesa* spp.), which suggests that rock glacial streams can act as refuge areas after the glacier loss [73]. However, further studies are necessary to demonstrate that cold-hardy *D. steinboecki* and other *Diamesa* species restricted to kryal habitat might survive competition with spring fauna (EPT) in rock glacier outflows.

4.5 Chemical pollution

Among the stressors that threaten the glacial biodiversity, there are also chemicals, i.e., persistent organics pollutants (POPs) deriving from long-range atmospheric transport and pesticides and emerging contaminants (e.g., personal care products as fragrances and polybrominated diphenyl ethers (PBDEs) widely used as flame retardants) carried to the glaciers by short-medium range atmospheric transport. These pollutants undergo cold condensation and accumulate in the glacier ice until their release in melt waters and ice-free soil [74, 75].

Among organic contaminants detected in glacier-fed streams, attention was paid to the insecticide chlorpyrifos, since high toxicity to insects and peak release by glacier melting occur concurrently with the period in which the streams are more densely populated by macroinvertebrates [76]. Chlorpyrifos and other organophosphate insecticides are known to exhibit increased toxicity in invertebrates at elevated temperatures [77]. Specifically, warming influences chlorpyrifos uptake in aquatic insects magnifying its negative effect on fauna. Other contaminants are heavy metals released by remains of the Great War, such as bombs, bullets, cannon parts and barbed wire buried in the ice 100 years ago, that are emerging due to glaciers retreating. These new sources of contamination have been recently documented for ice melt waters in the Italian Alps (mainly by nickel, arsenic and lead, unpublished data). Contamination of soils of the 1914–1918 Western Front zone, in Belgium and France by copper, lead and zinc was previously detected by [78]. Pesticides, fragrances and heavy metals affect swimming behavior and metabolism of *Diamesa* species from glacier-fed streams [79] at trace concentration (in order of ng/L), with still unknown effects on aquatic food web and on terrestrial fauna (*via* food web). To our knowledge, the understanding of final environmental fate of such pollutants is still scarce and fragmentary. Recently, evidences of microplastic bioaccumulation are given for freshwater amphipods from Svalbard glacier-fed streams [80] and for *Diamesa zernyi* larvae from the Amola Glacier-fed stream (Italy). No information is now available on their effects on ground beetle fauna.

5. Role of glacial biodiversity in the glacial-ecosystem functioning and services

Given the global change scenarios, it is opportune to analyze the role of glaciated areas in the context of persistent change from the physiographic point of view as well as the glacial biodiversity they host.

Standardized long-term monitoring, additional high-quality empirical studies on key organisms and landforms, and further development of analytical methods are of extreme importance in helping to quantify the extinction debt better and to more successfully enhance and protect glacial biodiversity.

Glacier shrinkage will alter hydrological regimes, sediment transport, and biogeochemical and contaminant fluxes from rivers to oceans, which will profoundly influence the natural environment and the ecosystem services that glacier-fed rivers provide to humans, particularly provision of water for agriculture, hydropower and consumption [4]. Biodiversity influences ecosystem functioning through changes in the amount of resource use or water self-depuration processes (regulating services) but is also a source of scientific and tourist attraction (cultural service).

Glacial biodiversity has an intrinsic value; most of the species are highly specialized to live in harsh environments, thus highly vulnerable to changes also of low intensity, have low dispersal ability and in most of the glacialized area of the world are endemic. Therefore, glaciated areas become territories with a collection of communities and species that mostly differ from those dominant in middle and low altitudes as well as differ between geographic areas. In addition, insects and other arthropods (e.g., spiders) living on the moraines, on the glaciers or flying on the glaciers (e.g., chironomids and stoneflies) act as additional source of food for some high-altitude mammals and bird species living at the edge of ice-related landforms [81, 82]. Therefore, glacial biodiversity is able to furnish an additional important naturalistic value of glacialized mountain regions.

Aquatic and terrestrial insects living in glacialized areas are trophically connected [83, 84], but not all insect groups react in the same way to the ongoing glacier retreat. Increasing glacial retreat differently affects ground-dwelling and aquatic insect taxa: ground beetles respond faster to glacier retreat than do non-biting midges, at least in species richness and species turnover patterns [84]. It depends on how fast habitat conditions change in relation to glacier retreat: the terrestrial environment changes faster than the aquatic environment. For instance, an increase of 0.6°C in summer temperatures approximately doubled the speed of colonization of the recently deglaciated terrains by ground-dwelling invertebrates [39]. The glacial stream lengthens but the physicochemical features and hydrological regime may not change for a long time, until the surface of the glacier is reduced to a few hectares and finally the environment becomes less extreme for life. As a result, as long as the environmental conditions remain extreme, the community in the glacier-fed stream is not affected.

Because even small ecosystem fragments like glaciers or other ice-related landforms have conservation value for insect biodiversity and ecosystem services, a better understanding and delineation of the species that need to be protected is also important. Funding of long-term research activities on habitat conservation in general, and specifically on insect science and taxonomy, is especially important to evaluate and mitigate future changes in insect communities, obtain reliable insect time series and discover species before they go extinct.

6. Conclusions

A recent large-scale study aimed to investigate trends in insect abundances over space and time has brought evidences about an average decline of terrestrial insect abundance (ca. 9%) per decade and an increase of freshwater insect abundance (ca. 11%) per decade. Both patterns were particularly strong in North America and some European regions [85], and the hypothesized drivers are land-use change and climate change.

Ground beetles and non-biting midges are among the animals best adapted to live at high altitudes, specifically to colonize the glaciers and surrounding terrestrial and aquatic habitats. They are present with few species adapted to low temperatures and food scarcity, factors that make these habitats extreme for life.

Spatio-temporal shift in insect communities in relation to the ongoing climate change is one of the most common patterns in mountain regions, but it is important to highlight that it will not affect all species equally [85, 86]. Aquatic and terrestrial insect communities seem to be differently affected by climate change. Firstly, because temperature variability is stronger on the ground with respect to water, aquatic insects are required to make smaller behavioral or physiological adjustments than terrestrial insects and aquatic habitats will be more buffered from climate warming [86]. In addition, terrestrial insects may have wider opportunities to survive in appropriate microclimate colonizing or surviving in ice-related microhabitat, thanks to the higher microhabitat heterogeneity on the ground with respect to the aquatic habitat.

The chapter provides a synthesis about the fascinating adaptations in morphology, behavior and physiology in terrestrial and aquatic species, on the species distribution in relation to the ice-related landform heterogeneity and on which species are threatened with extinction due to climate change and pollution. The future challenge will be to try to improve the knowledge of the glacial biodiversity in high-altitude and high-latitude areas notwithstanding the difficulty of accessing most of these areas.

Three research goals should be addressed in the near future: (i) increase the studies aimed at describing the glacial biodiversity; (ii) plan long-term monitoring projects in key areas and (iii) improve the scientific communication about the threats on high-altitude habitats. Thus, with this chapter, we are confident to inspire young researches to investigate the life in glacial ecosystems of the world before its possible disappearance.

Acknowledgements

We are grateful to the several colleagues and students that since more than 20 years are cooperating with us in collecting ground beetles and non-biting midges in glacialized areas of the world. We are also thankful to the National and Natural Parks that allowed the permission to monitor the insect populations of the glacial and periglacial environments.

Conflict of interest

The authors declare no conflict of interest.

Author details

Mauro Gobbi* and Valeria Lencioni
Section of Invertebrate Zoology and Hydrobiology, MUSE-Science Museum, Trento, Italy

*Address all correspondence to: mauro.gobbi@muse.it

References

[1] Díaz S, Settele J, Brondizio ES, Ngo HT, Guèze M, Agard J, editors. Summary for Policymakers of the Global Assessment Report on Biodiversity and Ecosystem Services of the Intergovernmental Science-Policy Platform on Biodiversity and Ecosystem Services. Bonn, Germany: IPBES Secretariat; 2019. Available from: https://ipbes.net/global-assessment [Accessed: 15 February 2020]

[2] Sánchez-Bayoa F, Wyckhuysbcd KAG. Worldwide decline of the entomofauna: A review of its drivers. Biological Conservation. 2019;**232**:8-27. DOI: 10.1016/j.biocon.2019.01.020

[3] Dangles O, Rabatel A, Kraemer M, et al. Ecosystem sentinels for climate change? Evidence of wetland cover changes over the last 30 years in the tropical Andes. PLoS One. 2017;**12**(5):e0175814. DOI: 10.1371/journal.pone.0175814

[4] Milner AM, Khamis K, Battin TJ, Brittain JE, Barrand NE, Füreder L, et al. Glacier shrinkage effects on downstream systems. Proceedings of the National Academy of Sciences. 2017;**114**(37):9770-9778. DOI: 10.1073/pnas.1619807114

[5] Adler C, Huggel C, Orlove B, et al. Climate change in the mountain cryosphere: Impacts and responses. Regional Environmental Change. 2019;**19**:1225-1228. DOI: 10.1007/s10113-019-01507-6

[6] Jacobsen D, Dangles O. Ecology of High Altitude Waters. Oxford University Press; 2017. p. 271

[7] Anesio AM, Laybourn-Parry J. Glaciers and ice sheets as a biome. Trends in Ecology & Evolution. 2012;**27**(4):219-225. DOI: 10.1016/j.tree.2011.09.012

[8] Takeuchi N, Kohshima S. A snow algal community on Tyndall glacier in the southern Patagonia Icefield, Chile. Arctic, Antarctic, and Alpine Research. 2004;**36**:92-99. DOI: 10.1657/1523-0430(2004)036[0092:ASACOT]2.0.CO;2

[9] Anesio AM, Lutz S, Chrismas NAM, et al. The microbiome of glaciers and ice sheets. npj Biofilms Microbiomes. 2017;**3**(10). DOI: 10.1038/s41522-017-0019-0

[10] Coulson SJ, Midgley NG. The role of glacier mice in the invertebrate colonisation of glacial surfaces: The moss balls of the Falljökull, Iceland. Polar Biology. 2012;**35**:1651-1658. DOI: 10.1007/s00300-012-1205-4

[11] Zawierucha K, Kolicka M, Takeuchi N, Kaczmarek Ł. What animals can live in cryoconite holes? A faunal review. Journal of Zoology. 2017;**295**(3):159-169. DOI: 10.1111/jzo.12195

[12] Ambrosini R, Musitelli F, Navarra F, et al. Diversity and assembling processes of bacterial communities in cryoconite holes of a Karakoram glacier. Microbial Ecology. 2017;**73**:827-837. DOI: 10.1007/s00248-016-0914-6

[13] Hagvar S. Primary succession in glacier forelands: How small animals conquer new land around melting glaciers. In: Young SS, Silvern SE, editors. International Perspective on Global Environmental Change. Rijeka, Croatia: INTECH; 2012. pp. 151-172. DOI: 10.5772/26536

[14] Millar CI, Westfall RD. Distribution and climatic relationships of the American pika (*Ochotona princeps*) in the Sierra Nevada and Western Great Basin, USA; periglacial landforms as refugia in warming climates. Arctic, Antarctic, and Alpine Research. 2010;**42**:76-88

[15] Millar CI, Westfall RD, Evenden A, Holmquist JG, Schmidt-Gengenbach J, Franklin RS, et al. Potential Climatic Refugia in Semi-arid, Temperate Mountains: Plant and Arthropod Assemblages Associated With Active Rock Glaciers, Talus Slopes, and Their Forefield Wetlands. Vol. 387. Sierra Nevada, California, USA. Quaternary International; 2013. pp. 106-121. DOI: 10.1016/j.quaint.2013.11.003

[16] Gobbi M, Ballarin F, Compostella C, Lencioni V, Seppi R, Tampucci D, et al. Physical and biological features of an active rock glacier of the Italian Alps. The Holocene. 2014;**24**:1624-1631. DOI: 10.1177/0959683614544050

[17] Gobbi M, Ballarin F, Brambilla M, Compostella C, Isaia M, Losapio G, et al. Life in harsh environments: Carabid and spider trait types and functional diversity on a debris-covered glacier and along its foreland. Ecological Entomology. 2017;**42**(6):838-848. DOI: 10.1111/een.12456

[18] Tampucci D, Azzoni RA, Marano G, Boracchi P, Citterio C, Compostella C, et al. Debris-covered glaciers as habitat for plant and arthropod species: Environmental framework and colonization patterns. Ecological Complexity. 2017;**32**(Part A):42-52. DOI: 10.1016/j.ecocom.2017.09.004

[19] Kotze D, Brandmayr P, Casale A, Dauffy-Richard E, Dekoninck W, Koivula M, et al. Forty years of carabid beetle research in Europe—From taxonomy, biology, ecology and population studies to bioindication, habitat assessment and conservation. ZooKeys. 2011;**100**:55-148. DOI: 10.3897/zookeys.100.1523

[20] Lorenz W. CarabCat: Global database of ground beetles. In: Roskov Y, Ower G, Orrell T, Nicolson D, Bailly N, Kirk PM, et al, editors. Species 2000 & ITIS Catalogue of Life. Naturalis, Leiden, the Netherlands: Species; 2020. Available from: www.catalogueoflife.org/col [Accessed 24 February 2020]

[21] Moret P. Altitudinal distribution, diversity and endemicity of Carabidae (Coleoptera) in the páramos of Ecuadorian Andes. Annales de la Société entomologique de France (N.S.). 2009;**45**(4):500-510. DOI: 10.1080/00379271.2009.10697632

[22] Casale A, Vigna Taglianti A. I coleotteri carabidi delle Alpi occidentali e centro-occidentali (Coleoptera, Carabidae). Biogeographia. 1992; XVI:331-399

[23] Casale A, Taglianti V. A. I coleotteri caraboidei delle Alpi e Prealpi centrali e orientali e loro significato biogeografico (Coleoptera, Caraboidea). Biogeographia. 2005;**XXVI**:129-201

[24] Cranston PS. Chironomids: From Genes to Ecosystems. Vol. 482. East Melbourne, Australia: CSIRO; 1995

[25] Ferrington LC. Global diversity of non-biting midges (Chironomidae; Insecta-Diptera) in freshwater. Hydrobiologia. 2008;**595**:447. DOI: 10.1007/s10750-007-9130-1

[26] Gereben BA. Co-occurrence and microhabitat distribution of six Nebria species (Coleoptera: Carabidae) in an alpine glacier retreat zone in the Alps, Austria. Arctic and Alpine Research. 1995;**27**(4):371-379. DOI: 10.1080/00040851.1995.12003134

[27] Bernasconi MG, Borgatti MS, Tognetti M, Valle B, Caccianiga C, Casarotto C, et al. Checklist ragionata della flora e degli artropodi (Coleoptera: Carabidae e Arachnida: Araneae) dei ghiacciai Centrale e Occidentale del Sorapiss (Dolomiti d'Ampezzo). Frammenti. 2019;**9**:49-65

[28] Gobbi M, Isaia M, De Bernardi F. Arthropod colonization of a debris-covered glacier. The

Holocene. 2011;**21**(2):343-349. DOI: 10.1177/0959683610374885

[29] Tenan S, Maffioletti C, Caccianiga M, Compostella C, Seppi R, Gobbi M. Hierarchical models for describing space-for-time variationsin insect population size and sex-ratio along a primary succession. Ecological Modelling. 2016;**329**:18-28. DOI: 10.1016/j.ecolmodel.2016.02.006

[30] Bråten AT, Flø D, Hågvar S, Hanssen O, Mong CE, Aakra K. Primary succession of surface active beetles and spiders in an alpine glacier foreland, central South Norway. Arctic, Antarctic, and Alpine Research. 2012;**44**(1):2-15. DOI: 10.1657/1938-4246-44.1.2

[31] Vater A, Matthews JA. Succession of pitfall-trapped insects and arachnids on eight Norwegian glacier forelands along an altitudinal gradient: Patterns and models. The Holocene. 2015;**25**(1):108-129. DOI: 10.1177/0959683614556374

[32] Hågvar S, Steen R, Flø D. Ecology of alpine carabid beetles (Coleoptera, Carabidae) in a Norwegian glacier foreland, with a special focus on claw wearing to indicate relative age. Norwegian Journal of Entomology. 2017;**64**:82-111

[33] Gereben-Krenn BA, Krenn H, Strodl MA. Initial colonization of new terrain in an alpine glacier foreland by Carabid beetles (Carabidae, Coleoptera). Arctic, Antarctic, and Alpine Research. 2011;**43**(3):397-403. DOI: 10.1657/1938-4246-43.3.397

[34] Kaufmann R. Invertebrate succession on alpine glacier foreland. Ecology. 2001;**82**:2261-2278. DOI: 10.1890/0012-9658(2001)082[2261,ISO AAG]2.0.CO;2

[35] Brambilla M, Gobbi M. A century of chasing the ice: Delayed colonisation of ice-free sites by ground beetles along glacier forelands in the Alps. Ecography. 2014;**37**:33-42. DOI: 10.1111/j.1600-0587.2013.00263.x

[36] Schlegel J, Riesen M. Environmental gradients and succession patterns of carabid beetles (Coleoptera: Carabidae) in an alpine glacier retreat zone. Journal of Insect Conservation. 2012;**16**:657-675. DOI: 10.1007/s10841-011-9448-x

[37] Moret P, Barragán Á, Moreno E, et al. When the ice has gone: Colonisation of equatorial glacier forelands by ground beetles (Coleoptera: Carabidae). Neotropical Entomology. 2020;**49**:213-226. DOI: 10.1007/s13744-019-00753-x

[38] Tampucci D, Gobbi M, Cabrini E, Compostella C, Marano G, Pantini P, et al. Plant and arthropod colonization of a glacier foreland in a peripheral mountain range. Biodiversity. 2015;**16**:213-223. DOI: 10.1080/14888386.2015.1117990

[39] Kaufmann R. Glacier foreland colonisation: Distinguishing between short-term and long-term effects of climate change. Oecologia. 2002;**130**:470-475. DOI: 10.1007/s00442-001-0815-2

[40] Tampucci D, Gobbi M, Marano G, Boracchi P, Boffa G, Ballarin F, et al. Ecology of active rock glaciers and surrounding landforms: Climate, soil, plants and arthropods. Boreas. 2017;**46**(2):185-198. DOI: 10.1111/bor.12219

[41] Pivetti S, Tampucci D, Caccianiga M, Compostella C, Gobbi M. Carabidofauna (Coleoptera: Carabidae) di un rock glacier in una catena Alpina periferica (Alpi Orobie, Bergamo). Bollettino del Museo Civico di Storia Naturale di Verona, Botanica Zoologia. 2016;**40**:3-9

[42] Růžička V, Šmilauer P, Mlejnek R. Colonization of subterranean habitats by spiders in Central Europe. International Journal of Speleology.

2013;**42**(2):133-140. DOI: 10.5038/1827-806X.42.2.5

[43] Vera A, Zuñiga-Reinoso A, Muñoz-Escobar C. Historical perspective on the distribution of *Andiperla willinki* "Patagonian dragon" (Plecoptera: Gripopterygidae). Revista Chilena de Entomología. 2012;**37**:87-93

[44] Koshima S. A novel cold-tolerant insect found in a Himalayan glacier. Nature. 1984;**310**:225-227. DOI: 10.1038/310225a0

[45] Takeuchi N, Kohshima S, Yoshimura Y, Seko K, Fujita K. Characteristics of cryoconite holes on a Himalayan glacier, Yala Glacier Central Nepal. Bulletin of Glaciological Research. 2000;**17**:51-59

[46] Füreder L. High alpine streams: Cold habitats for insect larvae. In: Margesin R, Schinner F, editors. Cold-Adapted Organisms. Berlin, Heidelberg: Springer; 1999. pp. 181-196. DOI: 10.1080/00040851.1995.12003134.

[47] Salerno F, Gambelli S, Viviano G, Thakuri S, Guyennon N, D'Agata C, et al. High alpine ponds shift upwards as average temperatures increase: A case study of the Ortles–Cevedale mountain group (southern Alps, Italy) over the last 50 years. Global and Planetary Change. 2014;**120**:81-91. DOI: 10.1016/j.gloplacha.2014.06.003

[48] Hamerlík L, Svitok M, Novikmec M, Očadlík M, Bitušík P. Local, among-site, and regional diversity patterns of benthic macroinvertebrates in high altitude waterbodies: Do ponds differ from lakes? Hydrobiologia. 2014;**723**: 41-52. DOI: 10.1007/s10750-013-1621-7

[49] Novikmec M, Veselská M, Bitušík P, Hamerlík L, Matúšová Z, Klementová BR, et al. Checklist of benthic macroinvertebrates of high altitude ponds of the Tatra Mountains (Central Europe) with new records of two species for Slovakia. Check List. 2015;**11**:1522. DOI: 10.15560/11.1.1522

[50] Sæther OA. Metriocnemus van der Wulp: Seven new species, revision of species, and new records (Diptera : Chironomidae). Annales de Limnologie - International Journal of Limnology. 1995;**31**(1):35-64. DOI: 10.1051/limn/1995002

[51] Frouz J, Matěna J, Ali A. Survival strategies of chironomids (Diptera: Chironomidae) living in temporary habitats: A review. European Journal of Entomology. 2003;**100**:459-465. DOI: 10.14411/eje.2003.069

[52] Lencioni V, Fattori D, Nardi G, Latella L. Spatio-temporal dynamic of invertebrate community in two ponds in a livestock alpine wetland. Aquatic Insects. 2019;**40**:30-52

[53] Rossaro B, Lencioni V, Boggero A, Marziali L. Chironomids from southern alpine running waters: Ecology, biogeography. Hydrobiologia. 2006;**562**:231-246. DOI: 10.1007/s10750-005-1813-x

[54] Jacobsen D, Dangles O, Andino P, Espinosa R, Hamerlìk L, Cadier E. Longitudinal zonation of macroinvertebrates in an Ecuadorian glacier-fed stream: Do tropical glacial systems fit the temperate model? Freshwater Biology. 2010;**55**(6):1234-1248. DOI: 10.1111/j.1365-2427.2009.02348.x

[55] Cadbury SL, Milner A, Hannah DM. Hydroecology of a New Zealand glacier-fed river: Linking longitudinal zonation of physical habitat and macroinvertebrate communities. Ecohydrology. 2010;**4**(4):520-531. DOI: 10.1002/eco.185

[56] Pizzolotto R, Gobbi M, Brandmayr P. Changes in ground beetle assemblages above and below the treeline of the Dolomites after almost

30 years (1980/2009). Ecology and Evolution. 2014;**4**(8):1284-1294. DOI: 10.1002/ece3.927

[57] Khamis K, Hannah DM, Brown LE, Tiberti R, Milner AM. The use of invertebrates as indicators of environmental change in alpine rivers and lakes. The Science of the Total Environment. 2014;**493**:1242-1254. DOI: 10.1016/j.scitotenv.2014.02.126

[58] Lencioni V. Glacial influence and stream macroinvertebrate biodiversity under climate change: Lessons from the southern Alps. Science of The Total Environment. 2018;**622-623**:563-575. DOI: 10.1016/j.scitotenv.2017.11.266

[59] Lencioni V, Jousson O, Guella G, Bernabò P. Cold adaptive potential of chironomids overwintering in a glacial stream. Physiological Entomology. 2015;**40**:43-53. DOI: 10.1111/phen.12084

[60] Moret P, Aráuz M, Gobbi M, Barragán A. Climate warming effects in the tropical Andes: First evidence for upslope shifts of Carabidae (Coleoptera) in Ecuador. Insect Conservation and Diversity. 2016;**9**:342-350. DOI: 10.1111/icad.12173

[61] Brandmayr P, Zetto BT. Comunità a coleotteri carabidi delle Dolomiti Sudorientali e delle Prealpi Carniche. Brandmayr P (editor) Zoocenosi e Paesaggio – I Le Dolomiti, Val di Fiemme-Pale di S. Martino. Studi Trentini di Scienze Naturali. Acta Biologica. 1988;**64**:125-250

[62] Robinson CT, Thompson C, Freestone M. Ecosystem development of streams lengthened by rapid glacial recession. Fundamental and Applied Limnology. 2014;**185**:235-246. DOI: 10.1127/fal/2014/0667

[63] Cauvy-Fraunié S, Espinosa R, Andino P, Jacobsen D, Dangles O. Invertebrate metacommunity structure and dynamics in an Andean glacial stream network facing climate change. PLoS One. 2015;**10**(8):e0136793. DOI: 10.1371/journal.pone.0136793

[64] Kavanaugh DH. Investigations on present climatic Refugia in North America through studies on the distributions of Carabid beetles: Concepts, methodology and prospectus. In: Erwin TL, Ball GE, Whitehead DR, Halpern AL, editors. Carabid Beetles. Dordrecht: Springer; 1979. pp. 369-381

[65] Slatyer RA, Schoville SD. Physiological limits along an elevational gradient in a radiation of montane ground beetles. PLoS One. 2016;**11**(4):e0151959. DOI: 10.1371/journal.pone.0151959

[66] Bernabò P, Viero G, Lencioni V. A long noncoding RNA acts as a post-transcriptional regulator of heat shock protein (HSP70) synthesis in the cold hardy *Diamesa tonsa* under heat shock. PLoS One. 2020;**15**(4):e0227172. DOI: 10.1371/journal.pone.0227172

[67] Brown LE, Khamis K, Wilkes M, Blaen P, Brittain JE, Carrivick JL, et al. Functional diversity and community assembly of river invertebrates show globally consistent responses to decreasing glacier cover. Nature Ecology and Evolution. 2018;**2**:325-333. DOI: 10.1038/s41559-017-0426-x

[68] Stewart JR, Lister AM, Barnes I, Dalén L. Refugia revisited: Individualistic responses of species in space and time. Proceedings of the Royal Society B: Biological Sciences. 2010;**277**:661-671. DOI: 10.1098/rspb.2009.1272

[69] Růžička V, Zacharda M. Arthropods of stony debris in the Krkonoše Mountains, Czech Republic. Arctic and Alpine Research. 1994;**26**(4):332-338. DOI: 10.1080/00040851.1994.12003077

[70] Lencioni V, Spitale D. Diversity and distribution of benthic and hyporheic fauna in different stream types on an

alpine glacial floodplain. Hydrobiologia. 2015;**751**:73-87. DOI: 10.1007/s10750-014-2172-2

[71] Hotaling S, Finn DS, Giersch JJ, Weisrock DW, Jacobsen D. Climate change and alpine stream biology: Progress, challenges, and opportunities for the future. Biological Reviews. 2017;**92**:2024-2045. DOI: 10.1111/brv.12319

[72] Lösch B, Tolotti M, Alber R. Permafrost und Gewässerökologie—Quellen undBäche mit Blockgletschereinfluss. Geo Alp. 2015;**12**:163-182

[73] Brighenti S, Tolotti M, Bruno MC, Wharton G, Pusch MT, Bertoldi W. Ecosystem shifts in alpine streams under glacier retreat and rock glacier thaw: A review. The Science of the Total Environment. 2019;**675**:542-559. DOI: 10.1016/j.scitotenv.2019.04.221

[74] Rizzi C, Finizio A, Maggi V, Villa S. Spatial–temporal analysis of currently used pesticides in alpine glaciers. Environmental Pollution. 2019;**248**:659-666. DOI: 10.1016/j.envpol.2019.02.067

[75] Steinlin C, Bogdal C, Luthi MP, Pavlova PA, Schwikowski M, Zennegg M, et al. A temperate alpine glacier as a reservoir of polychlorinated biphenyls: Model results of incorporation, transport, and release. Environmental Science & Technology. 2016;**50**:5572-5579. DOI: 10.1021/acs.est.5b05886

[76] Lencioni V, Bellamoli F, Bernabò P, Miari F, Scotti A. Response of *Diamesa* spp. (Diptera: Chironomidae) from alpine streams to emerging contaminants and pesticides. Journal of Limnology. 2018;**77**:131-140. DOI: 10.4081/jlimnol.2018.1802

[77] Harwood AD, You J, Lydy MJ. Temperature as a toxicity identification evaluation tool for pyrethroid insecticides: Toxicokinetic confirmation. Environmental Toxicology and Chemistry. 2009;**28**:1051-1058. DOI: 10.1897/08-291.1

[78] Bausinger E, Bonnaire J, Preuss J. Exposure assessment of a burning ground for chemicalammunition on the great war battlefields of Verdun. Science of the Total Environment. 2007;**382**:259-271. DOI: 10.1016/j.scitotenv.2007.04.029

[79] Villa S, Di Nica V, Pescatore T, Bellamoli F, Miari F, Finizio A, et al. Comparison of the behavioural effects of pharmaceuticals and pesticides on *Diamesa zernyi* larvae (Chironomidae). Environmental Pollution. 2018;**238**:130-139. DOI: 10.1016/j.envpol.2018.03.029

[80] Iannilli V, Pasquali V, Setini A, Corami F. First evidence of microplastics ingestion in benthic amphipods from Svalbard. Environmental Research. 2019;**179**(Pt A):108811. DOI: 10.1016/j.envres.2019.108811

[81] Resvold J. Perennial ice and snow-covered land as important ecosystems for birds and mammals. Journal of Biogeography. 2016;**43**:3-12. DOI: 10.1111/jbi.12609

[82] Brambilla M, Capelli F, Anderle M, Forti A, Bazzanella M, Masiero G, et al. Landscape-associated differences in fine-scale habitat selection modulate the potential impact of climate change on white-winged Snowfinch *Montifringilla nivalis*. Bird Study. 2018;**65**:525-532. DOI: 10.1080/00063657.2018.1561646

[83] Raso L, Sint D, Mayer R, Plangg S, Recheis T, Brunner S, et al. Intraguild predation in pioneer predator communities of alpine glacier forelands. Molecular Ecology. 2014;**23**(15):3744-3754

[84] Lencioni V, Gobbi M. Do carabids (Coleoptera: Carabidae) and chironomids (Diptera: Chironomidae)

exhibit similar diversity and distributional patterns along a spatio-temporal gradient on a glacier foreland. Journal of Limnology. 2018;**77**(s1):187-195. DOI: 10.4081/jlimnol.2018.1794

[85] Van Klink R, Bowler DE, Gongalsky KB, Swengel AB, Gentile A, Chase JM. Meta-analysis reveals declines in terrestrial but increases in freshwater insect abundances. Science. 2020;**368**:417-420. DOI: 10.1126/science.aax9931

[86] Shah AA, Dillon ME, Hotaling S, Woods HA. High elevation insect communities face shifting ecological and evolutionary landscapes. Current Opinion in Insect Science. 2020. in press. DOI: 10.1016/j.cois.2020.04.002

Chapter 9

Variations of Lys Glacier (Monte Rosa Massif, Italy) from the Little Ice Age to the Present from Historical and Remote Sensing Datasets

Fugazza Davide, Senese Antonella, Azzoni Roberto Sergio, D'Agata Carlo, Cat Berro Daniele, Mercalli Luca, Ventura Fabiano, Smiraglia Claudio and Diolaiuti Guglielmina Adele

Abstract

Alpine glaciers respond to climate imbalance by adjusting their mass and length. In turn, these changes modify the glacial and periglacial environment, leading to increased supraglacial debris cover, the development of glacial lakes and glacier fragmentation. In this research, we investigated the evolution of Lys Glacier (Monte Rosa Group), by studying length, area and volume changes, and evolution of its supraglacial debris cover and proglacial lakes by means of historical sources and high-resolution aerial and satellite orthophotos. Lys Glacier retreated almost continuously, by nearly 2 km, from its maximum Little Ice Age position. More recently, the glacier lost 11.91% of its area between 1975 and 2014 and underwent fragmentation in 2009. Over the same period, glacier fragmentation and tongue stagnation affected the formation and rapid growth of a series of ice-contact lakes and led to a non-linear debris cover evolution. The glacier was also subjected to strong volume losses, with more than 135 m thinning on the ablation tongue from 1991 to 2014. Analysis of the meteorological records (1927–present) from the closest weather station reveals a considerable increase in average annual temperatures by more than 1°C from the mean of 1971–1989 to the mean of 1990–2017.

Keywords: alpine glacier, terminus fluctuations, area change, volume change, debris cover, glacial lakes

1. Introduction

Observations show that the strongest influence of global climate change is recorded in alpine environments and glaciers are the first to be affected by global warming [1]. The response of a glacier to climate imbalance occurs initially through mass loss; eventually, the glacier adjusts to mass changes by changing its geometry, including area and length [2]. These geometric changes are also often accompanied by a shift in the glacier geomorphology, including an increase in the glacier debris cover, which can then decouple the glacier response from the temperature signal, and the appearance of thermokarst features such as kettles and ice-contact lakes [3], with possible impacts on glacier hazards downstream [4]. In fact, debris-covered glaciers and the expansion of supraglacial debris cover on debris-free glaciers are increasingly prominent features of the world's glaciated catchments in mountain regions including Himalaya [5], Karakoram [6, 7], Andes [8], Alps [9], Southern Alps of New Zealand [10], and Caucasus [11].

Over the past century, Italian glaciers have been shrinking at high rates [12–14]. Among these, Lys Glacier, located in the Monte Rosa massif (**Figure 1**), can be considered paradigmatic of the changes affecting alpine glaciers. Its response to increasing temperature and reduced accumulation is in fact similar to that of most glaciers in the Alps and elsewhere [14–17]: terminus retreat, area reduction and decreasing ice thickness. In addition, Lys Glacier has also recently shown other climate-related changes, including detachment of the main debris-covered tongue from the rest of the glacier body and possible separation of the two branches of the glacier tongue (**Figure 1a**); variations in supraglacial debris cover (both surface and thickness) on the glacier tongue, thermokarst features (e.g. kettle ponds) and processes (**Figure 1b**); and calving processes at the terminus where an ice-contact lake developed in the late 1990s (**Figure 1c**).

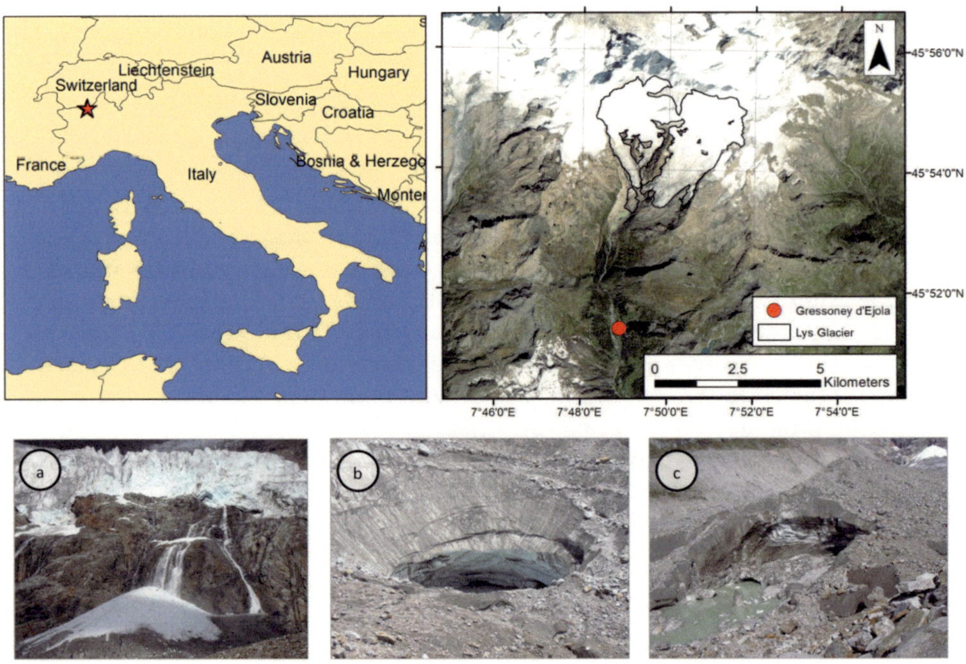

Figure 1.
Location of Lys glacier within Italy (red star) and of the weather station used in this study (red dot). The bottom pictures show some typical features found on the glacier tongue in recent years, including (a) separation of two tributaries from the main debris-covered tongue; (b) thermokarst features and processes; (c) calving processes at the ice-contact lake.

In the current context of temperature warming and glacier regression, long-term data reporting changes in glacier properties represent an important asset, as terminus, area and volume fluctuations can provide important information concerning the response of a glacier to climate imbalance. These data are a fundamental input for models that enable reconstructing past climate (from terminus fluctuations [18, 19]) and projecting future glacier changes [20], the availability of meltwater for domestic use and the production of hydroelectric energy [21, 22], as well as the formation of future lakes and potential hazards [23] and the impact of glacier change on tourism [24].

In this study, to describe the evolution of Lys Glacier, a multiple approach was followed. Terminus fluctuations since the early nineteenth century were analyzed using a variety of sources including detailed bulletins with reports of glaciological campaigns; over a more recent period (1975–2014), the area changes of the glacier were estimated by using remote sensing datasets, i.e. satellite and aerial orthophotos, while volume changes were evaluated by comparing a pair of digital elevation models (DEMs) obtained from cartography and satellite images. In addition, these sources permitted us to evaluate the evolution (i.e. surface cover and patterns) of the supraglacial debris cover and of the ice-contact lakes over the same period of observation and gain insights into the geomorphological evolution of the glacier.

2. Study site

Lys Glacier drains the southern flank of the Mont Rosa Group (45°54' N, 07°50′ E) (**Figure 1**). The most recent Italian glacier inventory ([14], data from 2005) reports a glacier area of 9.58 km^2, a south westerly aspect, an elevation range between 2392 and 4323 m a.s.l. and a length of 5.71 km. While Lys Glacier is presently the fourth largest Italian glacier, comparatively few studies have been conducted to investigate its evolution over the past century and recent decades. Strada [25] described terminus fluctuations until the early 1980s; Pelfini et al. [26] estimated the glacier response time using dendrochronology; Rota et al. [27] used DEMs from cartographic sources to calculate volume changes between 1925 and 1994. None of these studies or iconographic sources show evidence of continuous supraglacial debris cover until the late 1980s. Since then, debris cover has been present on the lower sector of the ablation tongue, initially as the continuation of medial moraines at an elevation below 2550 m a.s.l. The medial moraines formed below the glacier icefalls (**Figure 2**), thanks to the debris supplied by the surrounding deglaciated rock walls (owing to macrogelivation processes, permafrost degradation and structural rock falls). In more recent years, debris has come to cover the entire glacier tongue below 2550 m a.s.l., until its detachment from the upper sector of the glacier, which occurred in 2009 [28]. The occurrence of debris cover is thus a recent phenomenon and probably a consequence of the present deglaciation, which has affected Lys Glacier as well as the other alpine glaciers [1, 15]. Although Lys Glacier cannot be considered a debris-covered glacier even in recent times (as only a small sector of the glacier surface is continuously debris-covered, see [29]), the supraglacial morphologies and the processes affecting the debris-covered glacier tongue are comparable to those of larger debris-covered glaciers in Asia and in the Alps, including thermokarst processes and the formation of cavities and ice-contact lakes [29].

Figure 2.
Comparison of historical photographs of Lys glacier. (a) 1989. The glacier exhibits medial moraines on the distal part of its tongue, which is well developed. (b) 2019. The glacier has retreated above a rock wall in the western sector, leaving a dead ice tongue underneath, with a proglacial lake developing in the depression left by the disappearing tongue. The moraine ridges from the 1980s advance phase are also evident. (photo credits: (a) Willy Monterin-© Archivio Monterin. (b) Fabiano Ventura-© F. Ventura-sulletraccedeighiacciai.com).

3. Materials and methods

3.1 Terminus fluctuations

In this study, the terminus fluctuations of Lys Glacier were retrieved from different sources; early data concerning the nineteenth century are the same used in Strada [25] and references therein, i.e. early cartographic sources (map of the states of Sardinia from 1818 and cadastral maps), photographs and descriptions, particularly those in Monterin [30]; from the early twentieth century, length changes were obtained by extracting the field measurements available from the journals published by the Italian Glaciological Committee (CGI) [31, 32]. Surveys of the glacier terminus positions consist of tape measurements from fixed reference points in the glacier forefield, made at or near the end of the balance year, i.e. in the autumn season, with an estimated accuracy about 0.5 m, which can become worse in case of bad environmental conditions, poorly documented switches to new reference points, measurements taken over very long distances or at points on the snout outside the main flow line, or residual snow patches [12]. While the actual uncertainty from these issues is difficult to quantify, glacier terminus fluctuations remain an important asset to assess global climate and environmental change [19].

We followed recommendations by Citterio et al. [12] to perform quality-checking of the measurements extracted from CGI journals, to minimize the issues related to changing reference points or surveyors over the years. When multiple measurements were available from the same survey, an averaged variation was calculated. In case of missing years in the record, periods up to 5 years where the reference point had remained the same were filled by uniformly distributing the total terminus variation over the missing years.

In case of a gap associated with an undocumented change in a reference point, the gap was not filled, and the first measurement following the gap was considered as the starting point for comparison with the measurement resulting for the following year, as in Citterio et al. [12]. The data record for Lys Glacier from CGI journals is rather uninterrupted and permits analyzing the glacier history in detail over the last century.

3.2 Cartographic analysis

3.2.1 Data sources and preprocessing

Maps, aerial orthophotos and satellite images were analyzed to calculate the area and volume changes of Lys Glacier and changes in the area of the proglacial lakes. The available cartographic sources include large-scale maps from the Val d'Aosta region, produced in 1975 and 1991 at a nominal scale of 1:10000 (**Table 1**). The maps were available in digital form as rasters, projected in the UTM32N coordinate

Remote sensing data	Date	Resolution (m)	Usage
Regional technical map	1975	10	Glacier outlines
Aerial orthophoto	07/09/1988	0.5	Glacier outlines, debris cover mapping
Regional technical map	1991	10	Glacier outlines, DEM production, volume change
Aerial orthophoto	15/06/1994	0.5	Glacier outlines, debris cover mapping
Aerial orthophoto	31/08/1998	0.5	Glacier and lake outlines, debris cover mapping
Aerial orthophoto	12/09/2003	0.5	Glacier and lake outlines
Aerial orthophoto	04/09/2006	0.5	Glacier and lake outlines, debris cover mapping
Aerial orthophoto	16/07/2012	0.5	Glacier and lake outlines, debris cover mapping
Pleiades PHR1B	01/09/2014	2	Glacier outlines, DEM production, volume change, debris cover mapping

Table 1.
List of datasets used in this study and their usage; the acquisition date (whenever available) is reported as dd/mm/yyyy.

system based on the ED50 datum; contour lines and elevation points were digitized as shapefiles, and a DEM was produced from the 1991 map with a pixel spacing of 10 m using ArcMap Topo to Raster utility.

Aerial orthophotos from 1988 to 2012 (**Table 1**) were obtained from Geoportale Nazionale (http://www.pcn.minambiente.it/mattm/); these have a pixel size of 0.5 m and were downloaded in the UTM32 coordinate system based on the WGS84 datum. Most images were acquired in late summer months (August and September; **Table 1**), with minimum snow cover, while the orthophotos from 1988 and 2012 were acquired in early summer (June and July) and show evidence of residual snow, which however does not affect the glacier tongue. In addition to the aerial orthophotos, we obtained a stereo pair acquired from the Pleiades satellite constellation from the European Space Agency. The stereo pair was imaged on 1 September 2014 and provided at level 1B processing stage; we further processed it to obtain a DEM and an orthorectified image using NASA's AMES stereo pipeline (ASP). Eleven ground control points were selected from those available from Val d'Aosta regional authority (http://geonavsct.partout.it/pub/GeoNavITG/monografie.asp) to improve the geolocation accuracy of the DEM through bundle adjustment in the software; the DEM was then produced using the semi-global matching algorithm available in ASP [33], and the raw multispectral images were projected onto the DEM for orthorectification. Both the Pleiades DEM and multispectral image have a final resolution of 2 m.

3.2.2 Calculation of area and volume changes

Based on the available planar data (**Table 1**), the glacier and proglacial lake outlines were estimated over multiple years: in the 1975 and 1991 technical maps, they were already drawn on the map, while in the other datasets, they were manually digitized as part of this study. The spatial resolution of the orthophotos allowed us to clearly identify the outlines and to distinguish the debris-covered parts of the tongue from the proglacial areas. Before digitization, all available data were reprojected to the UTM32N coordinate system based on the WGS84 datum for consistency. We further evaluated the accuracy of the manual delineation using the buffer method proposed by Paul et al. [34], by allowing the glacier and lake outlines to grow and shrink with a buffer of 2 pixels.

The analysis of elevation and volume changes was based on the comparison of the DEMs from 1991 and 2014. Before performing this comparison, the DEMs were co-registered using the approach developed by Berthier et al. [35] and applied by Fugazza et al. [36]. In this approach, one DEM ('slave') is iteratively shifted with respect to a reference DEM ('master'), to minimize the standard deviation of elevation differences over stable areas located outside the glacier ($\sigma_{\Delta h}$). We selected the oldest DEM as the reference and resampled both DEMs to a common resolution of 10 m. By applying the co-registration, we obtained a residual $\sigma_{\Delta h}$ of 4.18 m; the average elevation difference over stable areas was 7 m, which was subtracted from the reference DEM. Based on the elevation difference on the glacier surface, we then calculated the volume change as:

$$\Delta V = \sum_{k=1}^{K} d_k A_k \tag{1}$$

where $k = 1 \ldots K$ are the pixels of the reference surface, A_k is the area of each pixel and d_k is the pixelwise difference between the 1991 and 2014 DEMs.

To express the uncertainty of volume changes, we used the approach described by Fischer et al. [37]: the uncertainty in elevation change ($\sigma_{\Delta h}$) is not considered as entirely correlated but scaled to account for the effective correlated area A_{cor}, based on the correlation length (i.e. the distance at which two pixels are correlated). We calculated the correlation length using an empirical semivariogram in R software as 200 m and the effective correlated area as 0.12 km^2. The uncertainty in volume change $\sigma_{\Delta h\_cor}$ is then expressed as:

$$\sigma_{\Delta h\_cor} = \sqrt{\sigma^2_{\Delta h} \cdot \frac{A_{cor}}{5 \cdot A_{1991}}} \quad (2)$$

where A_{1991} is the glacier area in 1991. The uncertainty calculation can in our case only be applied to the glacier tongue, as at altitudes above the glacier equilibrium line the presence of snow with little contrast deteriorates DEM reconstruction and leads to larger errors [38], particularly with the oldest DEM from 1991.

3.2.3 Mapping debris cover and its changes

To characterize the evolution of debris cover on Lys Glacier, we adopted the methodology proposed by Azzoni et al. [9] for glaciers in the Ortles-Cevedale Group. Supraglacial debris was mapped by employing a maximum likelihood (ML) supervised classification approach. Four classes were included in the classification: debris, ice, snow and shadow. Although we did not aim at mapping the other surface types, a higher number of classes was chosen to permit an improved mapping of debris cover; the shadow class was also included as some images are affected by topographic shadow; this occurred mostly in the accumulation basins except for the orthophoto from 2003, where shadows occur over the whole glacier; thus, this orthophoto was excluded from the analysis (**Table 1**). We also excluded the maps from 1975 and 1991, as the information on the presence of debris was not available. For each of the satellite/aerial orthophotos, we independently selected 10 to 15 training areas. To estimate the accuracy of the classification approach, we performed a validation test on the 2006 and 2012 orthophotos, by selecting 100 random points for each test and comparing the results of ML classification against those from a manual classification. The 2 years were chosen because they represent best (2006) and worst (2012) conditions for debris cover mapping, in terms of image quality, presence of snow and shadows. For both years, we calculated overall accuracy (the ratio of the number of correctly classified points to the number of total points), as well as producer's accuracy (PA) and user's accuracy (UA), for the different classes. PA denotes the ratio of the number of correctly classified points to the total (manually classified) point for a class and is the complement of omission errors; UA denotes the ratio of the number of correctly classified points to the number of predicted points for a class and is the complement of commission errors.

3.3 Meteorological data

In order to interpret the glaciological data of the study area, we analyzed meteorological variables recorded by the weather station installed in 1927 at Gressoney d'Ejola, 1850 m a.s.l., at a distance of 3.7 km from Lys Glacier (**Figure 1**). This long dataset allows investigating the recent climate behaviour, which influences the evolution of the glacier. At this weather station, manual observations of air temperature, liquid precipitation, total snow depth and thickness of fresh snow, atmospheric pressure, relative humidity, wind speed and velocity and cloud type and cover were collected three times a day (at 8 am, 2 pm and 7 pm) from 1927 to 2012. The data record is almost uninterrupted except for a short period from 1962 to 1970, when the station was temporarily moved downvalley, at 1730 m a.s.l. in the Orsia village. In 2002, an automatic weather station was installed at the same site, and it replaced the manual weather station after its dismissal in 2012. The automatic weather station measures air temperature, total precipitation (by a heated gauge) and snow height (by an automatic webcam) every hour. The contemporary presence of manual and automatic instruments has allowed ensuring the homogeneity and continuity of the series. Although a weather station also exists at a higher elevation and closer proximity to Lys Glacier (Alpe Courtlys, 1992 m a.s.l., 2.5 km distance), this station was installed in 2001 and does not permit a long-term analysis of climate variables. In this study, we calculated monthly, seasonal and annual averages of air temperatures and monthly, seasonal and annual cumulated liquid/solid precipitation for the analysis of climatological conditions.

4. Results

4.1 Terminus fluctuations

The history of the advances and retreats of Lys Glacier is summarized in **Figure 3**. We here report the complete record of data but also focus on the most recent period (1975–2017; see box in **Figure 3**). The general trend is one of retreat, after two short advance pulses during the Little Ice Age (LIA), culminating in 1822, when the glacier reached its peak LIA extension, and 1861. A very negative phase occurred between 1860 and 1882, when the glacier retreated by 941 m. This phase was followed by a short advance pulse at the end of the nineteenth century. The twentieth century saw an initial stability, with a small advance between 1913 and 1922, after which the glacier underwent an almost uninterrupted retreat, with the only exception of the short advance phase 1973–1985. The cumulative retreat since 1812 was 1.55 km, while since 1913 the glacier retreated by 847 m, with an average of -7.99 my$^{-1.}$

From 1975 to 2017, Lys Glacier retreated for 32 out of the 43 years analyzed, with an average variation of -8.26 my^{-1}. Since the last short-lived advance phase ending in 1985, retreat totalled 443 m (-13.85 my^{-1}). Retreat rates were particularly high between 2003 and 2007, with the most negative variations of −45 m (2007) and − 38 m (2003), while 1991 saw a very short readvance (+8 m), and in 2002 the glacier was considered stable. Retreat rates became lower after 2007, at -7.55 my^{-1} preluding to the separation of the glacier tongue from the parent glacier.

4.2 Area and volume changes

Over the period of observation, Lys Glacier underwent large changes in area. From 1975 to 1999, little change is evident on the western section of the glacier tongue (**Figure 4**), which reached the lowest elevations (ca. 2350 m a.s.l.). However, the easternmost part of the glacier tongue had already started retreating by 1999. Far greater changes than in the earlier period occurred from 1999 to 2014, with marked regression both in the western and eastern sections of the glacier tongue (**Figure 4a**). On the western section, the glacier tongue detached from the upper portion of the glacier at the base of an icefall, whereas on the eastern section the retreat exposed steep rocks underneath the ice. In 2014, the two main branches of the glacier still maintained a small connection at the base of a large rock outcrop (**Figure 4a**) and in the accumulation basins. Over the years, changes in Lys Glacier area were almost linear in rate, except for the first year, 1975, when the glacier had its largest area (11.3 km^2) of the period of observation (**Figure 5**). Since 1988, the area decreased at a rate of -0.045 km^{2}y^{-1}. In view of the high spatial resolution of the imagery, the uncertainty in the glacier outlines is below 2%, which supports the evidence of the general trend.

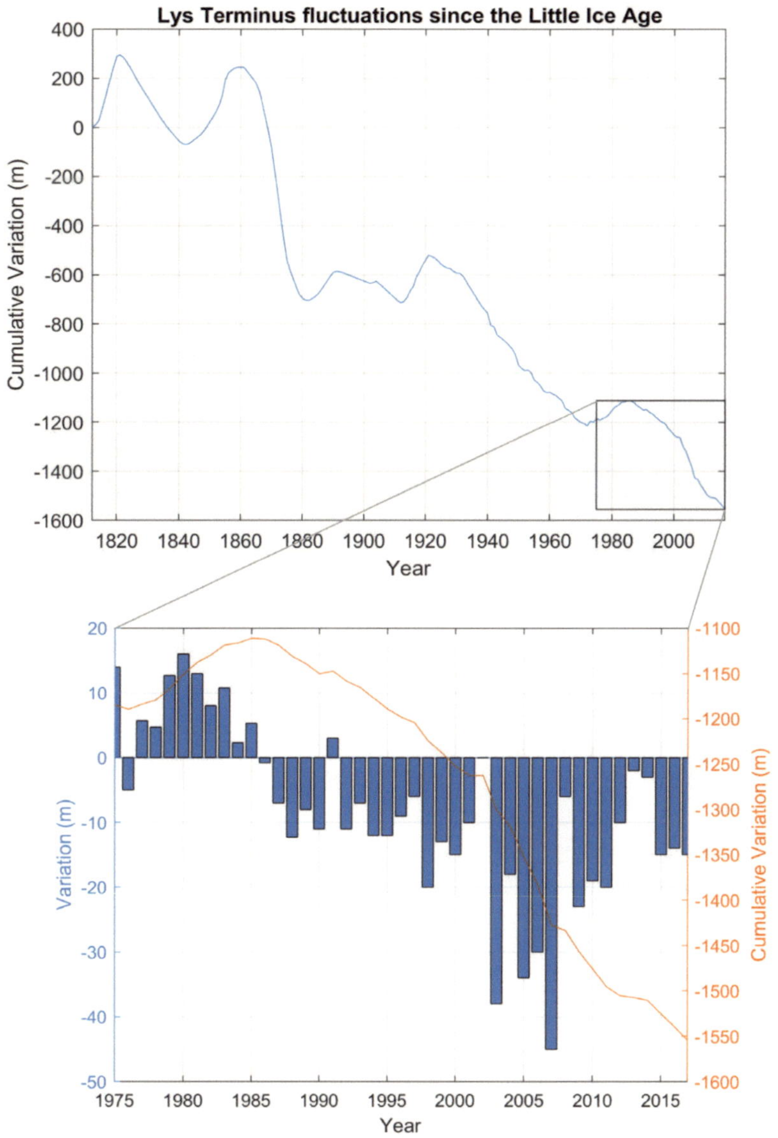

Figure 3.
Lys glacier terminus fluctuations. The box indicates the observation period analyzed in further detail, showing annual variations as well.

Beside changes in the glacier area, recent years have also seen a transformation of the geomorphological setting of the glacier outwash plain. While the exact date marking the origin of glacier lakes is difficult to establish, by 1999 five small ice-contact lakes had formed on the glacier tongue. Progressive stagnation in the ablation area allowed the lakes to grow and new ones to develop. Thus, from five scattered ponds in 1999 (totalling 1527 m^2) a larger overall area is seen together with the coalescence in two larger lakes in 2006 (8730 m^2 in total). In 2014, three further lakes formed while the existing ones grew in size (**Figure 4b, c**). The area of glacier lakes increased almost exponentially (summing the area of all lakes) from 1999, reaching 58,560 m^2 in 2014 (**Figure 5**). The uncertainty in the lake outlines is larger than that of the glacier outlines, up to 30% in 1999 because of their relatively small size.

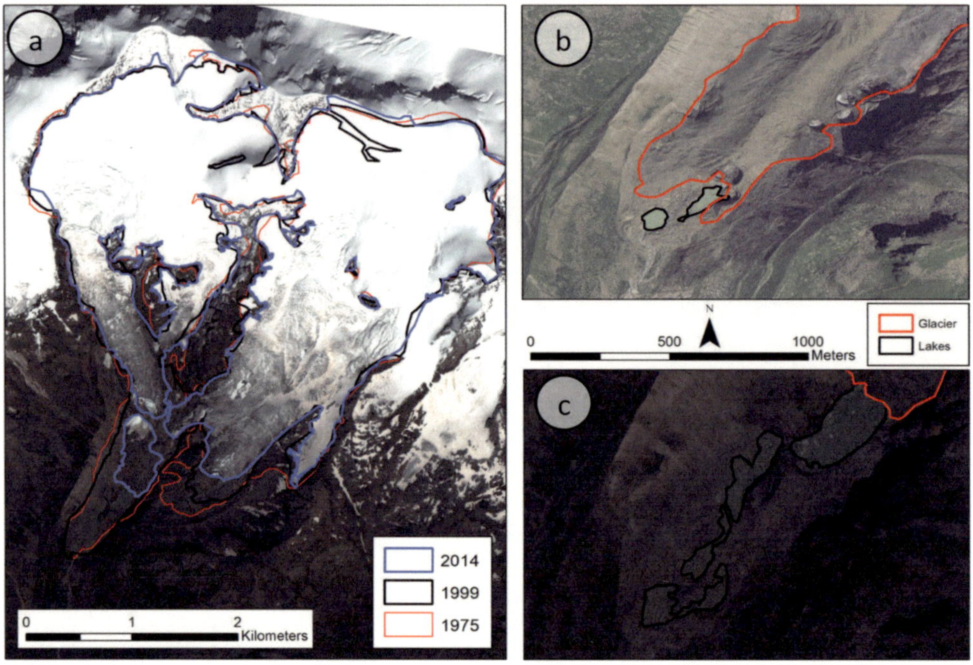

Figure 4.
Area changes of Lys glacier and its proglacial lakes over sample years. (a) Lys glacier outlines in 1975, 1999 and 2014. Background is the orthorectified Pleiades satellite image from 2014. (b) the glacier terminus and two proglacial lakes in 2006. (c) the glacier terminus and proglacial lakes in 2014.

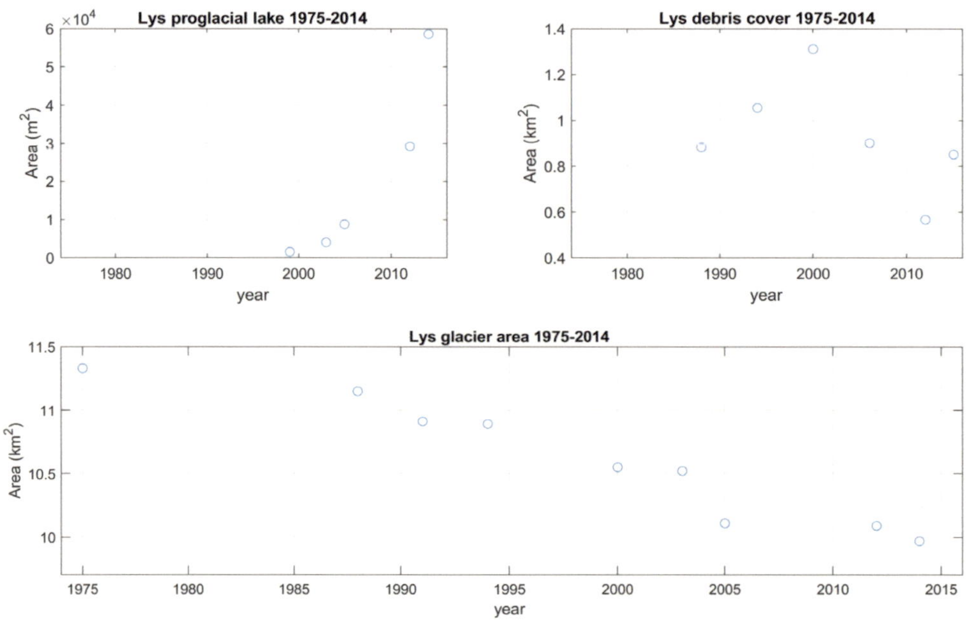

Figure 5.
Changes in the area of Lys glacier, its proglacial lake and supraglacial debris cover over the period of observation.

The DEM comparison shows a very different picture in the glacier tongue compared to the other areas of the glacier. On the tongue, large negative changes occurred, particularly on the western section, with a maximum of 136.52 m ice loss.

On the eastern part of the ablation tongue, the decrease in ice thickness reached instead 41 m. Aside from the glacier tongue, the other areas of the glacier show a more complex pattern, with areas of both positive and negative changes occurring. A loss in glacier thickness is seen particularly in the lower sectors of the western glacier tongue and in a narrow band in the uppermost reaches of the western accumulation basin, as well as on the eastern tongue in its lower easternmost area (**Figure 6**). A positive difference from 1991 to 2014 can instead be seen particularly on the western accumulation basin, where an apparent increase in thickness is

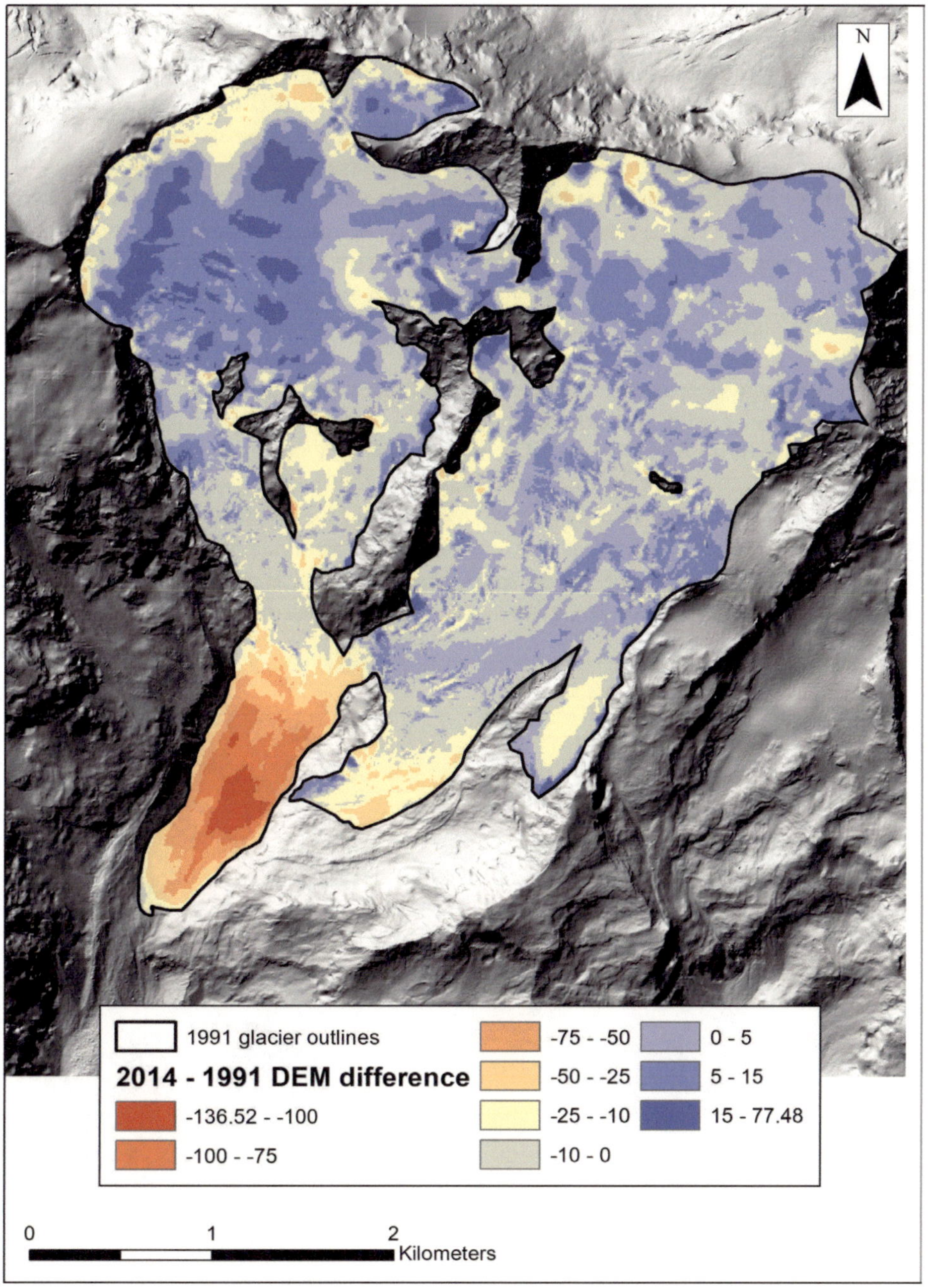

Figure 6.
Thickness changes of Lys glacier from 1991 to 2014. Negative values represent loss in ice thickness. The background image is the hillshade produced from the Pleiades DEM acquired in 2014.

observed between 10 and 20 m, and a maximum of above +75 m was recorded. Considering the entire glacier, the ice loss signal is still predominant, with an average of −4.34 m and a volume change of -47.06×10^6 m^3.

4.3 Debris cover evolution

On Lys Glacier, debris initially increased from 1988 to 2000, when it reached almost 1.3 km^2 on the glacier tongue (**Figures 5** and **7a**). In 1988, debris was present in narrow medial moraines and more abundantly at the terminus (**Figure 2**). In 2000, a more homogeneous coverage of the glacier tongue can be seen (**Figure 7b**), as a mantle of debris appears covering the glacier at the higher elevations at the margins of the western tongue, also suggesting a possible input from the lateral valley walls; conversely, on the eastern tongue and parts of the eastern accumulation basin, coverage is sparse and patchy. Since 2000, the expansion of supraglacial debris appears to have halted as the total area decreased (**Figure 5**): this effect was probably caused by the shrinking of the stagnating glacier tongue, which was however entirely debris-covered by 2005 (**Figure 7c** and **d**). The spread of supraglacial debris appears to have slowed down also at the higher elevations, although limited evidence for increasing coverage is seen for the eastern tongue of the glacier. The relatively high slope and the consequent presence of seracs might limit debris accumulation in those areas.

The accuracy of debris cover maps was evaluated separately for 2006 and 2012, as seen in **Tables 2** and **3**. In both years, PA and UA are very high for debris: both are 100% in 2006, while PA is 89% and UA 97% for 2012. Overall accuracy was 83% in 2006 and 71% in 2012. The limited overall accuracy in both tests is mostly caused by the difficulty in distinguishing between snow and ice (**Figure 7b**, where no ice was identified), which however is not crucial for the analysis of debris cover. We estimate the accuracy of debris cover mapping in other years to lie between the two values reported for 2006 and 2012; however, it is also possible that the debris cover

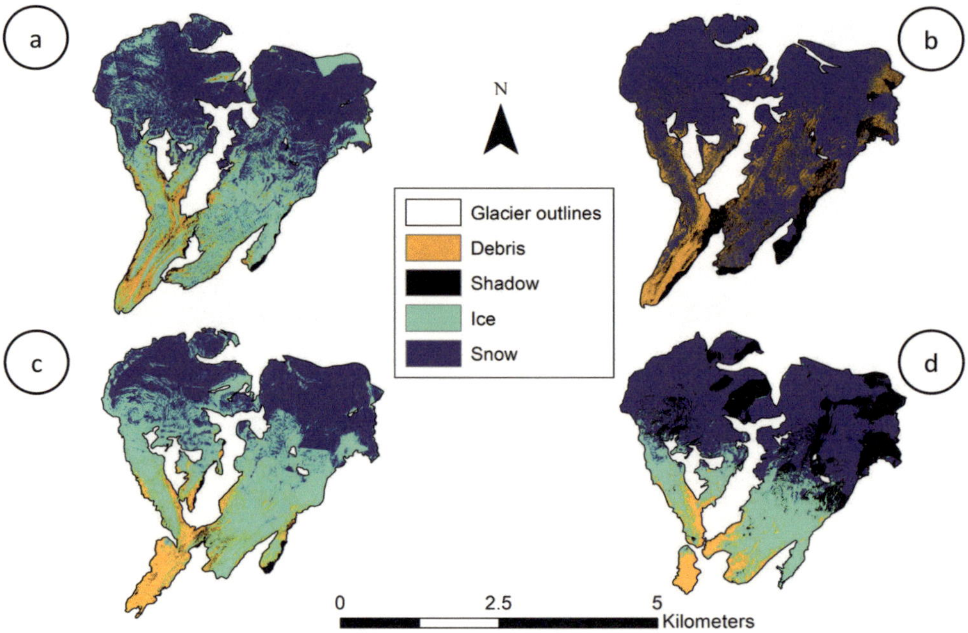

Figure 7.
Evolution of debris cover for Lys glacier. (a) 1988; (b) 2000; (c) 2005; (d) 2014.

		Manually classified				
		Debris	**Ice**	**Snow**	**Shadow**	**Totals**
Predicted	Debris	13	0	0	0	13
	Ice	0	21	17	0	38
	Snow	0	0	48	0	48
	Shadow	0	0	0	1	1
	Totals	13	21	65	1	100

Table 2.
Accuracy estimation for debris cover classification based on the aerial orthophoto from 2006.

		Manually classified				
		Debris	**Ice**	**Snow**	**Shadow**	**Totals**
Predicted	Debris	43	0	0	1	44
	Ice	5	8	23	0	36
	Snow	0	0	20	0	20
	Shadow	0	0	0	0	0
	Totals	48	8	43	1	100

Table 3.
Accuracy estimation for debris cover classification based on the aerial orthophoto from 2012.

amount after 2000 was underestimated, owing to the larger presence of snow cover and shadows in areas which were otherwise classified as debris-covered in previous years (compare **Figure 7b, c** and **d**).

4.4 Climatological analysis

Due to its high altitude and position, Lys valley experiences cold winters and temperate summers. Heavy rainfall can occur when south humid Mediterranean winds blow and collide against the orographic barrier of the southern slopes of Monte Rosa. Perturbations from the west and northwest are more frequent, but they discharge their rain/snow content mainly on the Mont Blanc and Valais areas, leaving the northeastern extremity of the Aosta Valley almost dry.

From 1928 to 2018 (excluding the period 1962–1970 when the station was temporarily moved downvalley), the mean annual temperature at Gressoney d'Ejola station was +4.4°C ranging from +2.6°C in 1984 to +6.1°C in 2015. Since we cannot focus on the commonly used 30-year reference period (1961–1990), we considered 1952–1961 and 1971–1990. Over this period of observation, the mean annual temperature was +4.0°C, slightly lower than the average of the past 30 years (1989–2018), which was +4.6°C. Climate warming is more evident when summer temperatures are compared: from +11.6°C during 1952–1961/1971–1990 to +12.7°C during 1989–2018, with a mean of +12.2°C over the whole period. The hottest month is July with a mean temperature of +13.2°C followed by August (+12.7°C) and June (+10.7°C). Generally, every 3 years the maximum daily of +25°C is recorded, even if the absolute maximum (+28.2°C) was observed on 4 August 2017 and 11 August 2003; 2003 was the hottest summer on record (average + 15.4°C; **Figure 8**) with relatively low amounts of liquid precipitation (216.6 mm corresponding to 74% of the mean summer total, 293.6 mm). Conversely, the coldest summer was 1977 (+9.9°C) with very heavy rainfall (428.6 mm corresponding to 146% of the mean summer total).

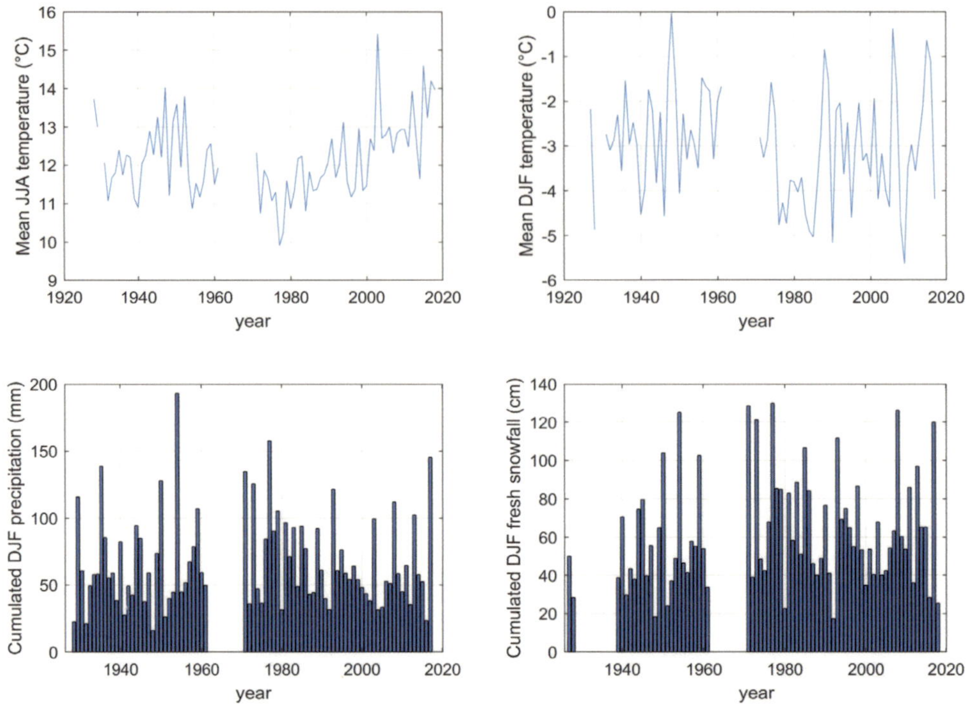

Figure 8.
Climatological analysis of the Gressoney d'Ejola weather station 1928–2018, including mean summer (JJA—June, July, August) temperature, mean winter (DJF—December, January, February) temperature, cumulative winter precipitation (liquid and solid) and cumulative winter fresh snowfall.

During the winter seasons, the monthly mean temperature is −2.6°C in December, −3.6°C in January and − 2.8°C in February with an absolute minimum up to −25.0°C recorded on 10 February 1986. The mean winter temperature is −3.0°C, and the coldest season was 2009–2010 with an average temperature of −5.6°C (and relatively low precipitation: 174.7 mm of total precipitation, 67.5 cm of snow depth, 181 cm of fresh snow; **Figure 8**), while the warmest one was 1948–1949 (0.0°C), also characterized by the lowest total precipitation (48.4 mm corresponding to 24% of the total mean winter amount, 201.4 mm; **Figure 8**), the lowest mean snow depth and cumulative fresh snow (7.7 cm and 52 cm corresponding to 12% of the mean winter amount—63.9 cm—and 28% of the mean winter total fresh snow, 188.7 cm, respectively) and the lowest number of days with snow cover (58 days corresponding to 67% of the mean total winter days, 86.8 days).

Generally, the coldest day is on 5 January (−4.5°C on average). Frost days ($T_{max} < 0°C$) generally occur from October to April, even if days with $T_{min} < 0°C$ can occur even in July. Thaw ($T_{ave} > 0°C$) begins at the end of June.

Comparing the two 30-year periods, the mean annual cumulated precipitation (liquid and solid) in 1952–1961/1971–1990 was 1126.9 mm, slightly higher than the amount of 1989–2018 (1090.7 mm). The same results can be observed looking at cumulative fresh snow: from 450.0 cm in 1952–1961/1971–1990 to 385.6 cm in 1989–2018 and from 201.4 cm to 187.5 cm when winter amount are considered. However, the variability remained the same: minima and maxima reached similar values during the two 30-year periods. The maximum total solid precipitation was recorded in winter 1954–1955 (579.7 mm, almost three times the mean value for winter) when very high values of mean snow depth (128.3 cm), cumulative fresh snow (376 cm; **Figure 8**) and number of days with snow cover (90 days) were observed. In this season, the temperature was equal to the average (−3.0°C).

Considering the whole year (average of 1927–2018), 1134.6 mm of rain and melted snow are distributed in 111 rainy and/or snowy days. Total precipitation is lower than the amount falling in the areas downvalley, which are even more exposed to the southerly humid winds, but much higher than the amount received in the dry Aosta Valley where 500 mm per year are generally recorded owing to its intra-alpine position. From November to April, there is an average 371.7 cm of fresh snowfall each year, almost equally divided between the various months: 61.9 cm per month ranging from 54.9 cm in November to 65.9 in December. Heavy snowfalls occur with Mediterranean temperate humid winds, with the daily maximum of 120 cm recorded on 1 January 1986 and the monthly maximum of 250 cm in April 1989.

5. Discussion

Since the LIA, and particularly over recent times, Lys Glacier has undergone remarkable changes in all the investigated parameters of length, area, volume and debris cover. The terminus fluctuation curve for Lys Glacier is strikingly similar to that of other published curves for Mer de Glace (Mont Blanc region, France) and Unterer Grindelwald (Bernese Oberland, Switzerland), which share some of the longest records of terminus fluctuations for alpine glaciers [39, 40], suggesting that in spite of the distance between these glaciers, the climatic setting with a predominant influence from westerly winds is similar. All three glaciers share two distinct advance phases during the LIA, interrupted by a period of retreat, which according to Vincent et al. [41] was caused by a decrease in winter precipitation; small differences however exist in the timing and magnitude of such advances: the maximum length of Lys Glacier in the past two centuries was reached in 1821, similarly to Mer de Glace (and Rosenlauigletscher, [40]), while the extent of Unterer Grindelwald and most other alpine glaciers peaked around the 1850s [40]. Well documented for all three glaciers is also the rapid retreat following the end of the LIA around 1860, although Mer de Glace also shows a small readvance around 1867, for which no evidence exists for Lys Glacier. Both Lys and Mer de Glace then enter a period of relative stability until the 1930s (although marked by several small advances and retreats); a period of marked retreat follows, attributed to enhanced solar radiation [42] lasting until the short-lived advance phase of the late 1960s/early 1970s, which lasted longer, up to 1995, for Mer de Glace compared to Lys and which is generally observed for mostly alpine glaciers [12], although smaller glaciers tend to have larger readvance periods in the twentieth century, indicating a shorter reaction time [43].

At Gressoney d'Ejola, the late 1970s appear particularly favorable years for glacialism, with cool summers and high amounts of winter snowfall. Since the 1980s, a clear warming trend has emerged for summer temperatures, which are 1.3°C higher from the mean of 1971–1989 to the mean of 1990–2017 (**Figure 8**) and 1.1°C higher when including 1952–1961/1971–1989 (+12.7°C in 1990–2017 compared to +11.6°C), while no clear signal can be seen in winter temperatures, total and solid precipitation. This is in line with trends previously observed for Italy [44] and high elevation regions [45] and explains the large retreat rates seen since 1985 (−13.85 my^{-1} in 1985–2017). Considering a longer set of temperature data analyzed for the Alps (1856–1998), the whole twentieth century was characterized by rising temperatures, at a rate of 0.50°C per century (considering summer temperatures [46]), even before the record-breaking decades of the 2000s and 2010s.

As concerns glacier area, the values reported here are always larger than those of the recent Italian glacier inventory (9.58 km^2 in 2005 [14]), even in 2014, because their outlines did not take into account a small area in the eastern part of the glacier, which was here included for consistency with the 1975 and 1991 outlines drawn on the maps. The rate of change of Lys Glacier found in this study (-0.44 kmy^{-1} or $-$ 0.4% y^{-1}) is however comparable to that of alpine glaciers, both in Italy and in the other alpine countries [14, 46, 47]. The increase in debris cover is lower than that reported by Azzoni et al. [9] for the Ortles-Cevedale region, where 38 glaciers were reported to have on average a 13.3% higher proportion of their area covered in debris in 2012 (reaching 30%) than 2003, while Lys Glacier went from 7.9% in 1988 to 12.4% in 2000, decreasing again to 8.5% in 2014. These values are also lower than those reported by Shukla et al. [5] for Samudratapu Glacier in Indian Himalayas, where debris cover nearly doubled over less than 3 years and in line with observations of glaciers in Caucasus by Stokes et al. [11], who describe a 3–6% increase in debris cover and a 57% increase in supraglacial and proglacial lake area. The formation of lakes on stagnating debris-covered glacier tongues was also observed by Kirkbride and Warren [48] and is attributed to the presence of ice-cored moraine which prevents meltwater runoff and favors the accumulation of water in depressions left by melting ice. Similarly to Stokes et al. [11], we also found that debris cover has not halted glacier retreat, counter to the evidence that a thick debris cover is known to reduce ablation (see, e.g. [49]). A field campaign conducted in 2006 revealed that debris thickness is generally above 10 cm and up to 60 cm, well above the critical threshold for which the insulating effect prevails on the albedo effect [49]. Thus, while it is possible that mass wasting would have been even higher without debris, it is more likely that glacier retreat occurred owing to the presence of ice-contact lakes and cavities, which enhance melt through backwasting [50].

Unlike retreat rates, mean thickness and volume changes for Lys Glacier are noticeably lower than in similar studies conducted on alpine glaciers: D'Agata et al. [21] report a decrease in ice thickness of -14.91 m for glaciers in Sondrio Province, Central Italian Alps, from 1981 to 2007, corresponding to -0.57 my^{-1}, while we estimated thinning of Lys Glacier to be 0.19 my^{-1}. In their study of all glaciers in the Swiss Alps, Fischer et al. [37] report an area-weighted mass balance of -0.62 m w.e. y^{-1} from 1980 to 2010, while the geodetic mass balance of Lys Glacier (using a conversion factor of 0.85 accounting for the average density of ice and firn as done by [37]) would be -0.16 m w.e. y^{-1}, which is at the low end of the scale for glaciers analyzed in the study by Fischer et al. [37], although a few Swiss glaciers do share a similar geodetic mass balance. The geodetic mass balance of Lys Glacier would also be lower than that reported by Berthier et al. [51], i.e. -1.05 ± 0.37 m w.e. y^{-1} for glaciers in the Mont Blanc massif using ASTER, SPOT and Pleiades DEMs between 2000 and 2014. If we exclude the possibility that debris cover has slowed down thinning for reasons stated above, the large differences and the apparent low average elevation changes of Lys Glacier can be explained by (1) the relatively large size of the glacier accumulation basins and the presence of seasonal snow on the Pleiades image (**Figure 2a**) and (2) interpolation errors, especially in the oldest cartographic DEM, also seen particularly affecting the accumulation basins of the glacier where little change is expected to occur over the years [38], while we observe areas with apparent thickening of 20 m and up to 75 m. Considering the glacier tongue, the glacier thinning is however evident: to compare our findings against those of Rota et al. [27], we selected an area below 2660 m a.s.l., corresponding to 0.7 km^2. The average ice loss was computed as 62.92 ± 0.81 m from 1991 to 2014, equal to a thinning rate of 2.74 ± 0.03 my^{-1}, and higher than the values reported by Rota et al. [27], i.e. -1.32 my^{-1} between 1925 and 1953 and $-$ 0.42 my^{-1} between 1953 and 1994, suggesting an increase in the glacier tongue thinning rates. Our values for the glacier tongue are also in line with the findings of Mölg and Bolch [52] who reported an average elevation change of -67 ± 5.3 my^{-1} for Zmuttgletscher (Swiss Alps), albeit for a larger area and a longer period, between 1946 and 2005.

6. Conclusion

In this study, we analyzed the evolution of Lys Glacier, one of the largest glaciers in the Italian Alps, by looking at a variety of parameters: terminus fluctuations were studied from historical sources and glaciological bulletins from 1812 to 2017; changes in surface, debris cover and area of supraglacial/proglacial lakes together with volume changes were examined from cartographic and remote sensing datasets from 1975 to 2014. The glacier length variations were found to be similar to those of large glaciers in the Alps such as Mer de Glace and Unterer Grindelwald, indicating a similar climatic setting in spite of the distance of these glaciers; the worst conditions for the glacier development occurred after the end of the Little Ice Age and since 1985 (−443 m from 1985 to 2017) reflecting increasing temperatures as seen from the closest weather station located at Gressoney d'Ejola. Overall, Lys Glacier has retreated by almost 1.6 km since the LIA.

All the other glaciological findings point to a strong glacier reduction, which is interpreted as an evident impact of climate change: the rate of area change was $-0.04\ km^2\ y^{-1}$ since 1988, while glacier volume decreased by $-47 \times 10^6\ m^3$ from 1991 to 2014. The glacier debris cover increased from 1988 to 2000, when it covered 12.4% of the glacier area, and then started decreasing again, as a result of glacier shrinking, while the area of the proglacial lakes grew exponentially over the same period. We consider the changes in area and debris cover as highly reliable in view of our accuracy assessment (max 2% error in the glacier outlines and accuracy between 90% and 100% when mapping debris cover), while the uncertainty in volume variations is larger because of the lower quality of the input DEM from 1991.

In view of the present conditions of the glacier, which prevent reaching the glacier tongue, remote sensing remains as the only viable option to investigate the glacier variations in the future, while the detachment of the glacier tongue has further complicated studying terminus fluctuations. To further our understanding of the glacier past conditions, other historical sources should be considered, including pictorial documents to lengthen the record of glacier terminus position and aerial photography from the past century to provide more accurate estimates of volume changes.

Acknowledgements

We acknowledge the DAR—Department of Regional Autonomies—of the Italian presidency of the Council of Ministers for funding this research.

The photographic comparison of Lys Glacier was performed by Fabiano Ventura in the context of the project 'On the Trail of the Glaciers'— sulletraccedeighiacciai.com.

Pleiades images were obtained from the European Space Agency (project ID 32535). We wish to thank Val d'Aosta region for providing access to the GPS reports and regional technical maps; the meteorological series from Gressoney d'Ejola is available thanks to the efforts and dedication of Umberto and Willy Monterin.

Conflict of interest

The authors declare no conflict of interest.

Author details

Fugazza Davide[1*], Senese Antonella[1], Azzoni Roberto Sergio[1], D'Agata Carlo[1], Cat Berro Daniele[2], Mercalli Luca[2], Ventura Fabiano[3], Smiraglia Claudio[1] and Diolaiuti Guglielmina Adele[1]

1 Department of Environmental Science and Policy, Università degli studi di Milano, Milano, Italy

2 Società Meteorologica Italiana, Moncalieri (TO), Italy

3 Associazione MacroMicro, Roma, Italy

*Address all correspondence to: davide.fugazza@unimi.it

References

[1] Zemp M, Frey H, Gärtner-Roer I, Nussbaumer SU, Hoelzle M, Paul F, et al. Historically unprecedented global glacier decline in the early 21st century. Journal of Glaciology. 2015;**61**(228):745-762

[2] Cuffey K, Paterson WSB. The Physics of Glaciers. 4th ed. Butterworth-Heinemann/Elsevier: Burlington, MA; 2010. p. 693

[3] Diolaiuti G, Smiraglia C. Changing glaciers in a changing climate: How vanishing geomorphosites have been driving deep changes in mountain landscapes and environments. Géomorphologie: Relief, Processes, Environment. 2010;**16**(2):131-152

[4] Keiler M, Knight J, Harrison S. Climate change and geomorphological hazards in the eastern European Alps. Philosophical Transactions of the Royal Society A-Mathematical Physical and Engineering Sciences. 2010;**368**(1919):2461-2479

[5] Shukla A, Gupta RP, Arora MK. Estimation of debris cover and its temporal variation using optical satellite sensor data: A case study in Chenab basin, Himalaya. Journal of Glaciology. 2009;**55**(191):444-452

[6] Minora U, Bocchiola D, D'Agata C, Maragno D, Mayer C, Lambrecht A, et al. Glacier area stability in the Central Karakoram National Park (Pakistan) in 2001-2010: The "Karakoram anomaly" in the spotlight. Progress in Physical Geography: Earth and Environment. 2016;**40**(5):629-660

[7] Senese A, Maragno D, Fugazza D, Soncini A, D'Agata C, Azzoni RS, et al. Inventory of glaciers and glacial lakes of the Central Karakoram National Park (CKNP – Pakistan). Journal of Maps. 2018;**14**(2):189-198

[8] Janke JR, Bellisario AC, Ferrando FA. Classification of debris-covered glaciers and rock glaciers in the Andes of Central Chile. Geomorphology. 2015;**241**:98-121

[9] Azzoni RS, Fugazza D, Zerboni A, Senese A, D'Agata C, Maragno D, et al. Evaluating high-resolution remote sensing data for reconstructing the recent evolution of supra glacial debris: A study in the Central Alps (Stelvio Park, Italy). Progress in Physical Geography: Earth and Environment. 2018;**42**(1):3-23

[10] Kirkbride MP. The temporal significance of transitions from melting to calving termini at glaciers in the central southern Alps of New Zealand. The Holocene. 1993;**3**(3):232-240

[11] Stokes CR, Popovnin V, Aleynikov A, Gurney SD, Shahgedanova M. Recent glacier retreat in the Caucasus Mountains, Russia, and associated increase in supraglacial debris cover and supra–/proglacial lake development. Annals of Glaciology. 2007;**46**:195-203

[12] Citterio M, Diolaiuti G, Smiraglia C, D'agata C, Carnielli T, Stella G, et al. The fluctuations of Italian glaciers during the last century: A contribution to knowledge about alpine glacier changes. Geografiska Annaler. Series A, Physical Geography. 2007;**89**(3):167-184

[13] D'Agata C, Bocchiola D, Maragno D, Smiraglia C, Diolaiuti GA. Glacier shrinkage driven by climate change during half a century (1954-2007) in the Ortles-Cevedale group (Stelvio National Park, Lombardy, Italian Alps). Theoretical and Applied Climatology. 2014;**116**(1-2):169-190

[14] Smiraglia P, Azzoni RS, D'Agata C, Maragno D, Fugazza D, Diolaiuti GA. The evolution of the Italian glaciers from the previous data base to the new Italian inventory. Preliminary considerations and results. Geografia Fisica e Dinamica Quaternaria. 2015;**38**:79-87

[15] Paul F, Frey H, Le Bris R. A new glacier inventory for the European Alps from Landsat TM scenes of 2003: Challenges and results. Annals of Glaciology. 2011;**52**(59):144-152

[16] Nuimura T, Sakai A, Taniguchi K, Nagai H, Lamsal D, Tsutaki S, et al. The GAMDAM glacier inventory: A quality-controlled inventory of Asian glaciers. The Cryosphere. 2015;**9**(3):849-864

[17] Seehaus T, Malz P, Sommer C, Lippl S, Cochachin A, Braun M. Changes of the tropical glaciers throughout Peru between 2000 and 2016–Mass balance and area fluctuations. The Cryosphere. 2019;**13**(10):2537-2556

[18] Mackintosh AN, Anderson BM, Pierrehumbert RT. Reconstructing climate from glaciers. Annual Review of Earth and Planetary Sciences. 2017;**45**(1):649-680

[19] Goosse H, Barriat P-Y, Dalaiden Q, Klein F, Marzeion B, Maussion F, et al. Testing the consistency between changes in simulated climate and alpine glacier length over the past millennium. Climate of the Past. 2018;**14**:1119

[20] Zekollari H, Huss M, Farinotti D. Modelling the future evolution of glaciers in the European Alps under the EURO-CORDEX RCM ensemble. The Cryosphere. 2019;**13**(4):1125-1146

[21] D'Agata C, Bocchiola D, Soncini A, Maragno D, Smiraglia C, Diolaiuti GA. Recent area and volume loss of Alpine glaciers in the Adda River of Italy and their contribution to hydropower production. Cold Regions Science and Technology. 2018;**148**:172-184

[22] Schaefli B, Manso P, Fischer M, Huss M, Farinotti D. The role of glacier retreat for Swiss hydropower production. Renewable Energy. 2019;**132**:615-627

[23] Frey H, Haeberli W, Linsbauer A, Huggel C, Paul F. A multi-level strategy for anticipating future glacier lake formation and associated hazard potentials. Natural Hazards and Earth System Sciences. 2010;**10**(2):339-352

[24] Wang S-J, Zhou L-Y. Integrated impacts of climate change on glacier tourism. Advances in Climate Change Research. 2019;**10**(2):71-79

[25] Strada E. Le variazioni del ghiacciaio del Lys dalla «Piccola glaciazione» ai nostri giorni. Natura Bresciana. 1987;**24**:275-288

[26] Pelfini M, Belloni S, Rossi G, Strumia G. Response time of the Lys glacier (Valle d'Aosta). An example of dendrogeomorphological and environmental study. Geografia Fisica e Dinamica Quaternaria. 1997;**20**:329-338

[27] Rota T, Bonacina E, De Lotto P, Rossi G, Smiraglia C. Variazioni volumetriche della lingua del ghiacciaio del Lys (Monte Rosa, Valle d'Aosta) nel XX secolo. Supplementi Geografia Fisica e Dinamica Quaternaria. 2001;**5**:183-185.

[28] Comitato Glaciologico Italiano M. Report of the glaciological survey 2009. Baroni C, Meneghel M, Mortara G, editors. Geografia Fisica e Dinamica Quaternaria. 2010;**33**(2):68

[29] Kirkbride MP. Debris-covered glaciers. In: Singh VP, Singh P, Haritashya UK, editors. Encyclopedia of Snow, Ice and Glaciers. Berlin: Springer; 2011. pp. 190-192. Available from: https://discovery.dundee.ac.uk/en/publications/debris-covered-glaciers

[30] Monterin U. Le variazioni secolari del clima del Gran San Bernardo: 1818-1931 e le oscillazioni del Ghiacciaio del Lys al Monte Rosa: 1789-1931. Bollettino del Comitato Glaciologico Italiano. 1932;**12**(1):59-188

[31] Comitato Glaciologico Italiano. Campagne Glaciologiche. Bollettino del Comitato Glaciologico Italiano. 1914-1977;**I-II**:1-25

[32] Comitato Glaciologico Italiano. Campagne Glaciologiche. Geografia Fisica e Dinamica Quaternaria. 2018;**1978**:1-41

[33] Beyer RA, Alexandrov O, McMichael S. The Ames stereo pipeline: NASA's open source software for deriving and processing terrain data. Earth and Space Science. 2018;**5**(9):537-548

[34] Paul F, Barrand NE, Baumann S, Berthier E, Bolch T, Casey K, et al. On the accuracy of glacier outlines derived from remote-sensing data. Annals of Glaciology. 2013;**54**(63):171-182

[35] Berthier E, Arnaud Y, Kumar R, Ahmad S, Wagnon P, Chevallier P. Remote sensing estimates of glacier mass balances in the Himachal Pradesh (Western Himalaya, India). Remote Sensing of Environment. 2007;**108**(3):327-338

[36] Fugazza D, Scaioni M, Corti M, D'Agata C, Azzoni RS, Cernuschi M, et al. Combination of UAV and terrestrial photogrammetry to assess rapid glacier evolution and map glacier hazards. Natural Hazards and Earth System Sciences. 2018;**18**(4):1055-1071

[37] Fischer M, Huss M, Hoelzle M. Surface elevation and mass changes of all Swiss glaciers 1980-2010. The Cryosphere. 2015;**9**(2):525-540

[38] Ragettli S, Bolch T, Pellicciotti F. Heterogeneous glacier thinning patterns over the last 40 years in Langtang Himal, Napal. The Cryosphere. 2016;**10**(5):2075-2097

[39] Nussbaumer SU, Zumbühl HJ, Steiner D. Fluctuations of the "Mer de glace" (Mont Blanc area, France) AD 1500-2050: An interdisciplinary approach using new historical data and neural network simulations. Zeitschrift für Gletscherkunde und Glazialgeologie. 2007;**40**:1-183

[40] Zumbühl HJ, Nussbaumer SU. Little ice age glacier history of the central and western Alps from pictorial documents. Cuadernos de Investigación Geográfica. 2018;**44**(1):115-136

[41] Vincent C, Meur EL, Six D, Funk M. Solving the paradox of the end of the little ice age in the Alps. Geophysical Research Letters. L09706. 2005;**32**(9):1-4. Available from: https://agupubs.onlinelibrary.wiley.com/doi/abs/10.1029/2005GL022552

[42] Nussbaumer SU, Steinhilber F, Trachsel M, Breitenmoser P, Beer J, Blass A, et al. Alpine climate during the Holocene: A comparison between records of glaciers, lake sediments and solar activity. Journal of Quaternary Science. 2011;**26**(7):703-713

[43] Nussbaumer SU, Zumbühl HJ. The little ice age history of the glacier des Bossons (Mont Blanc massif, France): A new high-resolution glacier length curve based on historical documents. Climatic Change. 2012;**111**(2):301-334

[44] Brunetti M, Maugeri M, Monti F, Nanni T. Temperature and precipitation variability in Italy in the last two centuries from homogenised instrumental time series. International Journal of Climatology. 2006;**26**(3):345-381

[45] Pepin N, Bradley RS, Diaz HF, Baraer M, Caceres EB, Forsythe N, et al. Elevation-dependent warming in mountain regions of the world. Nature Climate Change. 2015;**5**(5):424-430

[46] Fischer M, Huss M, Barboux C, Hoelzle M. The new Swiss glacier

inventory SGI2010: Relevance of using high-resolution source data in areas dominated by very small glaciers. Arctic, Antarctic, and Alpine Research. 2014;**46**(4):933-945

[47] Gardent M, Rabatel A, Dedieu J-P, Deline P. Multitemporal glacier inventory of the French Alps from the late 1960s to the late 2000s. Global and Planetary Change. 2014;**120**:24-37

[48] Kirkbride MP, Warren CR. Tasman glacier, New Zealand: 20th-century thinning and predicted calving retreat. Global and Planetary Change. 1999;**22**(1):11-28

[49] Mihalcea C, Mayer C, Diolaiuti G, Lambrecht A, Smiraglia C, Tartari G. Ice ablation and meteorological conditions on the debris-covered area of Baltoro glacier, Karakoram, Pakistan. Annals of Glaciology. 2006;**43**:292-300

[50] Steiner JF, Pellicciotti F, Buri P, Miles ES, Immerzeel WW, Reid TD. Modelling ice-cliff backwasting on a debris-covered glacier in the Nepalese Himalaya. Journal of Glaciology. 2015;**61**(229):889-907

[51] Berthier E, Cabot V, Vincent C, Six D. Decadal region-wide and glacier-wide mass balances derived from multi-temporal ASTER satellite digital elevation models. Validation over the Mont-Blanc area. Frontiers in Earth Science. 2016;**4**:63. Available from: http://journal.frontiersin.org/Article/10.3389/feart.2016.00063/abstract

[52] Mölg N, Bolch T. Structure-from-motion using historical aerial images to analyse changes in glacier surface elevation. Remote Sensing. 2017;**9**(10):1021